T0386135

Flora of Tropical East Africa - Eriocaulaceae (1997)

Editor

R.M. Polhill, B.A., Ph.D., F.L.S.

CRC Press is an imprint of the
Taylor & Francis Group, an **informa** business
A BALKEMA BOOK

CRC Press
Taylor & Francis Group
6000 Broken Sound Parkway NW, Suite 300
Boca Raton, FL 33487-2742

© 1997 by Taylor & Francis Group, LLC
CRC Press is an imprint of Taylor & Francis Group, an Informa business

No claim to original U.S. Government works

ISBN 13: 978-90-6191-377-1 (hbk)

This book contains information obtained from authentic and highly regarded sources. Reasonable efforts have been made to publish reliable data and information, but the author and publisher cannot assume responsibility for the validity of all materials or the consequences of their use. The authors and publishers have attempted to trace the copyright holders of all material reproduced in this publication and apologize to copyright holders if permission to publish in this form has not been obtained. If any copyright material has not been acknowledged please write and let us know so we may rectify in any future reprint.

Except as permitted under U.S. Copyright Law, no part of this book may be reprinted, reproduced, transmitted, or utilized in any form by any electronic, mechanical, or other means, now known or hereafter invented, including photocopying, microfilming, and recording, or in any information storage or retrieval system, without written permission from the publishers.

For permission to photocopy or use material electronically from this work, please access www.copyright.com (http://www.copyright.com/) or contact the Copyright Clearance Center, Inc. (CCC), 222 Rosewood Drive, Danvers, MA 01923, 978-750-8400. CCC is a not-for-profit organization that provides licenses and registration for a variety of users. For organizations that have been granted a photocopy license by the CCC, a separate system of payment has been arranged.

Trademark Notice: Product or corporate names may be trademarks or registered trademarks, and are used only for identification and explanation without intent to infringe.

Visit the Taylor & Francis Web site at
http://www.taylorandfrancis.com

and the CRC Press Web site at
http://www.crcpress.com

FLORA OF TROPICAL EAST AFRICA

ERIOCAULACEAE

SYLVIA PHILLIPS

Annual or perennial herbs usually under 1 m. high, often much smaller; stem usually abbreviated to a basal disc, infrequently elongate. Leaves narrow, lanceolate to filiform, spirally arranged, crowded into a basal rosette or rarely dispersed on elongate stems, opaque to translucent, sometimes fenestrate. Inflorescence capitate, single or in umbels, on leafless ribbed scapes. Capitula composed of many small (often tiny) unisexual flowers on a central receptacle surrounded by 1–several whorls of involucral bracts, monoecious, each flower often subtended by a floral bract. Flowers trimerous or less often dimerous, subactinomorphic to strongly zygomorphic. Perianth usually composed of two distinct whorls, sometimes reduced or absent; calyx of free, partially or almost completely connate sepals, sometimes spathe-like especially in the male flowers; petals free or partially connate, spathulate to filiform, frequently hairy, often with a subapical black gland; male petals often fused with the floral axis to form an infundibular structure topped by very small free lobes. Male flowers with stamens as many or twice as many as the sepals; inner whorl epipetalous; a vestigial gynoecium usually present. Female flowers with a superior 2–3-locular ovary; style simple, tipped with 2, 3 or 6 elongate stigmas. Fruit a thin-walled, loculicidally dehiscent capsule, each locule containing a single seed.

10 genera throughout the tropics and subtropics in marshy or seasonally inundated places, especially in montane areas on sandy soils, rarely truly aquatic; a few species of *Eriocaulon* also in temperate areas. The greatest concentration of genera and species is found in upland parts of South America; *Eriocaulon* is the only genus in Asia.

The family is easily recognisable by its capitula of small crowded flowers, atop leafless scapes and surrounded by involucral bracts. The capitula are reminiscent of *Compositae* but, unlike that family, are never brightly coloured, occurring in shades of white, dirty grey, brown or black.

Stamens twice as many as the petals, 6 or 4; petals usually with a black subapical gland:
  Petals of female flowers free; leaves, scapes and sheaths usually glabrous . . . . 1. **Eriocaulon**
  Petals of female flowers connate into a tube except at the base; leaves, scapes and sheaths usually hairy . . . . 2. **Mesanthemum**
Stamens as many as the petals, 3; petals usually eglandular:
  Petals of female flowers connate in the middle, the bases and tips free; leaves, scapes and sheaths usually hairy and often glandular . . . . 3. **Syngonanthus**
  Petals of female flowers free:
    Leaves, scapes and sheaths glabrous; petals of male flowers with distinct free lobes . . . . 1. **Eriocaulon**
    Leaves, scapes and sheaths hairy; petals of male flowers connate into a truncate tube lacking free lobes . . . . 4. **Paepalanthus**

1. **ERIOCAULON**

L., Sp. Pl.: 87 (1753) & Gen. Pl., ed. 5: 38 (1754); N.E. Br. in F.T.A. 8: 231–259 (1901); Ruhland in E.P. 4, 30: 30–117 (1903); H. Hess in Ber. Schweiz. Bot. Ges. 65: 123–178 (1955); Meikle in F.W.T.A., ed. 2, 3: 57–64 (1968); Oberm. in F.S.A. 4(2): 9–19 (1985)

Scapes single, often twisted, enclosed at the base by a tubular sheath. Involucral bracts of the capitulum scarious to coriaceous, spreading or reflexed at maturity; floral bracts frequently white-hairy towards the tip. Flowers sessile or pedicellate on the central receptacle, male and female mixed or the female around the periphery (especially in annuals). Sepals free or ± connate especially in the male flowers, usually free in the female, often navicular, sometimes winged. Petals with a subapical or apical black gland on the inner face, or the glands sometimes reduced or absent, often white-hairy at the tip; male petals all very small or one enlarged and exserted from the capitulum. Stamens twice as many as the petals (except *E. angustibracteum*); anthers black or white. Seeds relatively large, ellipsoid, yellow to reddish brown, smooth or distinctively patterned.

About 400 species worldwide in tropical and subtropical regions; one species in temperate eastern U.S.A. and western Europe. The exact number of species is very uncertain, but speciation is well developed throughout the whole range of the genus.

Identification of most *Eriocaulon* species is not easy due to the very small size of the flowers, which nevertheless exhibit a wide range of important differences. Frequently species of extremely similar external appearance have widely different floral morphology. Only with a few of the most distinctive species is it safe to name a specimen without first dissecting the capitulum and examining the flowers. This particularly applies to the small ephemerals, often with black glabrous capitula, which spring up around the margins of drying pools and runoffs. Several species often grow together in these locations and care must be taken to avoid mixed gatherings.

The flowers easily hydrate in a drop of water for dissection, without the need for boiling. Measurements of floral parts are taken from hydrated flowers. Other measurements refer to dried material; leaves and scapes often shrink considerably on drying. It is important where possible to examine mature capitula, as the dimensions of the floral parts can alter drastically at maturity; in particular in strongly zygomorphic flowers the longest petals elongate rapidly at anthesis in the male flowers and when seed is ripe in the female, and also wings on the sepals increase in size. Leaves usually taper to some extent from the sheathing base to the tip and width measurements are taken about halfway up. Various types of hairs occur on the floral parts, and short, thick, obtuse, very opaque white hairs are widespread. Parts bearing such hairs are described here as white-papillose when the hairs are short, or white-pilose or white-villous when they are longer.

The patterning of the seed coat is of great importance in *Eriocaulon*, both for identification and for establishing relationships. Sometimes species with almost identical floral morphology have very different seeds. Seed-set can be sparse in the robust perennials, but otherwise seed is usually abundantly produced and is available for inspection. Although a high magnification is needed to see fine detail, the coarse patterning can be discerned with a hand lens, and this is often sufficient to confirm an identification.

Plant with a long stem clothed in numerous filiform leaves; submerged aquatic, only the capitula emergent · · · · · · 1. *E. setaceum*
Plant tufted, the leaves in a basal rosette; marsh plants or ruderals:
  Anthers white; petals absent in female flowers; sepals in female flowers 0–3, filiform, caducous; slender annual 2. *E. cinereum*
  Anthers black:
    Petals absent in female flowers; sepals in female flowers 2, linear-falcate (fig. 1/8) · 3. *E. polhillii*
    Petals present in female flowers:
      Flowers with 2 petals, 4 anthers and 2 stigmas:
        Scapes conspicuously papillose; capitula 1–2 mm. diameter · 4. *E. glandulosum*

Scapes smooth; capitula 1.5–4.5 mm. diameter:

Seeds uniformly brown; sepals of female flowers narrowly to very broadly winged (fig. 3/7) · · 5. *E. mutatum*

Seeds brown with a white-reticulate patterning (fig. 4/2); sepals of female flowers with a wing not wider than the sepal-body · · · · · · · · · · · 6. *E. nigrocapitatum*

Flowers with 3 petals, (3–)6 anthers and 3 stigmas:

Flowers with 2 sepals and 3 petals; capitula shiny white; small annual with longitudinally ridged seeds · 7. *E. truncatum*

Flowers with 3 sepals and 3 petals (2 broadly winged sepals in female flowers of *E. strictum*):

Perennials from a short rhizome, often robust capitula 8–20 mm. diameter, densely white-hairy:

Leaves not usually exceeding 20 cm. long, often much shorter, 1–5 mm. wide; scapes 4–8-ribbed; capitula 8–12 mm. diameter:

Sepals of female flowers connate into a spathe leaves acute · 8. *E. pictum*

Sepals of female flowers free; leaves tipped by a pore · 9. *E. teusczii*

Leaves large, thick, up to 40 cm. long and 4–18 mm. wide; scapes stout, 8–10-ribbed; capitula 9–20 mm. diameter:

Receptacle villous; female sepals villous around keel with long hyaline hairs; involucral bracts clearly shorter than capitulum width, neatly imbricate · · · · · 10. *E. iringense*

Receptacle glabrous or thinly and inconspicuously pilose; female sepals lacking long hyaline hairs; involucral bracts as wide as capitulum or only slightly shorter:

Female sepals glabrous inside, sometimes connate and spathe-like (fig. 2/9); receptacle glabrous; involucral bracts slightly shorter than capitulum width, coriaceous at least towards base · · · · · 11. *E. schimperi*

Female sepals bearded inside (fig. 2/8); receptacle pilose; involucral bracts as wide as capitulum, scarious · · · · · · · · · · · 12. *E. mesanthemoides*

Small ephemerals, annuals or occasionally perennial (but then capitula < 7 mm. diameter or leaves < 8 cm. long); capitula 1.5–9 mm. diameter, glabrous or hairy:

Capitula subsessile, the scapes not exceeding the leaf-rosette; dwarf montane species from above 3000 m. · 13. *E. volkensii*

Capitula rising above the leaves on well developed scapes:

Sepals of female flowers subequal, similar · · · Key 2

Sepals of female flowers very unequal, of different form, the two laterals with a distinct spongy or winged keel; median sepal much narrower (absent in *E. strictum*) · · · · Key 3

KEY 2

Capitula woolly with fluffy, long, hyaline hairs; involucral bracts villous with tubercle-based hyaline hairs (fig. 1/2) 14. *E. laniceps*
Capitula glabrous or pilose with short, stout, white hairs; involucral bracts glabrous:
Involucral bracts thin, soon crumpling and not visible in mature capitulum; scapes 4-winged; capitula black with a covering of short white hairs ··········· 15. *E. elegantulum*
Involucral bracts clearly visible in mature capitulum; scape-ribs rounded:
Sepals and petals of female flowers linear to filiform (fig. 3/8); male sepals free; capitula 5–8 mm. diameter, shaggy, pale and shining ··········· 16. *E. bongense*
Sepals of female flowers narrowly lanceolate or oblong-ovate; male sepals usually connate, at least near the base:
Male calyx acutely 3-lobed from ± the middle; seeds uniformly brown; small ephemerals up to 12 cm. high:
Capitula dark grey to blackish; floral bracts obovate, acute ··········· 17. *E. abyssinicum*
Capitula whitish to pale grey; floral bracts narrowly lanceolate, acuminate-aristate ··········· 18. *E. welwitschii*
Male calyx-lobes obtuse to truncate-denticulate, often almost completely connate; seeds with a white reticulate or striate patterning:
Sepals of female flowers with a spongy wing or thickening on the keel, usually hairy along the margins inside with long hyaline hairs; perennials from a short rhizome:
Capitula subglobose, slightly wider than long, 5.5–7 mm. diameter, sometimes viviparous; floral bracts oblanceolate-oblong, acute, the inner densely white-hairy; female sepals gibbous with a broad spongy wing wider than the sepal-body ··········· 19. *E. zambesiense*
Capitula globose to dome-shaped, often slightly longer than wide, 4–5 mm. diameter, never viviparous; floral bracts broadly spathulate-cuspidate, sparsely white-hairy; female sepals narrowly winged or merely thickened, wing narrower than sepal-body 20. *E. inyangense*
Sepals of female flowers sometimes winged but not conspicuously spongily thickened, the margins glabrous:
Involucral bracts scarious, reflexing at maturity:
Seeds transversely striate; sepals of female flowers slightly thickened to broadly winged, acute 21. *E. transvaalicum*
Seeds reticulate; sepals of female flowers usually unwinged, truncate-denticulate ··········· 22. *E. selousii*
Involucral bracts firm, not reflexing at maturity:
Capitula glabrous; leaves subulate to filiform, setaceous-tipped; seeds white-striate ··········· 23. *E. stenophyllum*
Capitula shortly white-hairy; leaves linear, acute; seeds white-reticulate ··········· 24. *E. afzelianum*

KEY 3

Floral bracts and flowers white-pilose; male sepals almost free — 25. *E. crassiusculum*
Floral bracts and flowers glabrous; male sepals connate at least below middle:
Floral bracts acute to acuminate, usually contrasting with involucral bracts; capitula 3.5–7 mm. wide:
Seed with obvious transverse rows of white papillae; all petals of female flowers eglandular:
Leaves narrowly linear to subulate, 3–10 cm. long, acuminate or setaceous-tipped; lateral sepals of female flowers with acuminate, often recurving tips; anthers usually 3 . . . . 26. *E. angustibracteum*
Leaves broadly linear, 1–3 cm. long, subacute (fig. 3/1); lateral sepals of female flowers with shortly acute tips; anthers 6 . . . . 27. *E. buchananii*
Seed ± smooth, faintly white-reticulate; lateral petals of female flowers glandular . . . . 28. *E. burttii*
Floral bracts rounded to subacute, ± concolorous with involucral bracts; capitula 2.5–4 mm. wide:
Sepals of male flowers free to below the middle; lateral sepals of female flowers falcate . . . . 29. *E. modicum*
Sepals of male flowers connate to their tips; lateral sepals of female flowers deeply concave, strongly gibbous:
Leaves linear-lanceolate, up to 1.5 cm. long, papillose; seed with white reticulate patterning; median sepal of female flowers usually present . . . . 30. *E. maculatum*
Leaves narrowly linear, up to 3 cm. long, smooth; seed transversely white-papillose; median sepal of female flowers absent . . . . 31. *E. strictum*

1. **E. setaceum** *L.*, Sp. Pl.: 87 (1753); Ruhland in E.P. 4, 30: 89, fig. 9 (1903); Meikle in F.W.T.A., ed. 2, 3: 62, fig. 337/9 (1968); Oberm. in F.S.A. 4(2): 10, fig. 2/1 (1985). Lectotype, chosen by Trimen in J.L.S. 24: 136 (1888): Sri Lanka, *Herb. Hermann* species no. 50 (BM, lecto.)

Aquatic; stem floating below the surface, slender, elongate, branching, up to 50 cm. or more long, densely clothed in numerous leaves, terminating in a stiff umbel of 1–20 scapes emergent above the water surface. Leaves filiform, yellow-green, flaccid, 3–10 cm. long, 0.1–0.3 mm. wide, 1-nerved, fenestrate, not sheathing at the base. Scapes up to 30 cm. high, 5–7-ribbed; sheaths shorter than the leaves, loose, parallel-sided, shortly 1–3-fid at the mouth. Capitulum depressed-globose, (2–)3–5(–6) mm. diameter, blackish or white-pubescent; involucral bracts shorter than the capitulum width, blackish, scarious, obovate-oblong to suborbicular, broadly rounded, weakly reflexing at maturity; floral bracts cuneate-oblong, blackish, concave, acute, the inner sometimes white-papillose on the back; receptacle glabrous or thinly pilose; flowers trimerous, 0.9–1.4 mm. long, pedicellate, black. Male flowers: sepals oblong, concave, unequally connate into a spathe or almost free, glabrous or white papillose near the rounded tip; petals small, glandular; anthers 6, black. Female flowers: sepals subequal or the median smaller, deeply concave, the back gibbous with broadly rounded tip, spongy along the midline, infrequently lightly keeled and winged with a few white hairs on the upper keel and an acute tip, upper margins smooth or denticulate; petals subequal, linear-spathulate, rather spongy, glands small or absent, glabrous or a few white papillae at the tip; ovary black. Capsule only very tardily dehiscent. Seeds 0.4–0.5 mm. long, transversely white-papillose, mid-brown but often appearing blackish due to the adherent ovary-wall.

UGANDA. Lango District: Orumo [Oruma], Sept. 1935, *Eggeling* 2214!; Teso District: Soroti, Omunyal swamp, 14 Sept. 1954, *Lind* 362!

TANZANIA. Mpanda District: 3 km. on Ugalla R. road from Mpanda–Uvinza road, 19 May 1997, *Bidgood et al.* 4039!; Njombe District: ± 1.5 km. beyond Igosi on Njombe–Kipengere road, 26 Apr. 1970, *Wingfield* 594!; Songea District: Kwamponjore valley, 26 Apr. 1956, *Milne-Redhead & Taylor* 9929!

DISTR. **U** 1, 3; **T** 4, 7, 8; widespread throughout the tropics, including Australia

HAB. Rooting in the mud of slow-moving or stagnant water up to 50 cm. deep, only the capitula emerging above the surface; 1000–1200 m.

SYN. *E. melanocephalum* Kunth, Enum. Pl. 3: 549 (1841); Ruhland in E.P. 4, 30: 89 (1903). Type: Brazil, São Paulo, *Sello* 5850 (B, holo., K, iso.!)

E. intermedium Körn. in Linnaea 27: 601 (1856). Types: India, *Wight* 2369 (B, syn., K, isosyn.) & Sri Lanka, *Thwaites* 791 (B, syn.)

E. bifistulosum Van Heurck & Müll. Arg. in Van Heurck, Obs. Bot.: 105 (1870); N.E. Br. in F.T.A. 8: 239 (1901); Ruhland in E.P. 4, 30: 90 (1903); H. Hess in Ber. Schweiz. Bot. Ges. 65: 130 (1955). Type: Nigeria, Nupe, *Barter* 1021 (AWH, G, K!, iso.)

E. fluitans Baker in J.L.S. 20: 277 (1883). Types: Madagascar, *Baron* 926 & *Parker* (K, syn.!)

E. capillus-naiadis Hook.f., Fl. Brit. Ind. 6: 572 (1893). Type: Bengal, *Griffith* (CAL, lecto., K, isolecto.!)

E. limosum Engl. & Ruhland in E.J. 27: 74 (1899). Type: Nigeria, Nupe, *Barter* 1021 (B, holo.), *non Barter* 1021 (AWH, G)

E. schweinfurthii Engl. & Ruhland in E.J. 27: 74 (1899). Type: Sudan, Dar-Fertit, Biri, *Schweinfurth* 224 (B, holo., K, iso.!)

NOTE. *E. setaceum* is immediately recognisable by its distinctive aquatic habit, the long floating stems clothed in numerous filiform leaves and terminated by an emergent cluster of often black or whitish capitula. Unlike the normal situation in *Eriocaulon*, where the mature seed is readily expelled from the capitulum by the dehiscence of the capsule, in *E. setaceum* the whole fruit is shed, sometimes splitting into 3 mericarps which fall separately. The seed thus appears blackish due to the adherent ovary-wall. The seed, surrounded by a layer of mucilage, is later expelled from this adherent wall through the usual loculicidal slit.

E. setaceum s.l. is one of the most widespread species of *Eriocaulon*, and comprises a complex of forms. Variation in the structure of the bracts and flowers has led to the description of several species throughout its range. These differences relate mainly to the colour of the capitulum; hairiness of the receptacle (glabrous or pilose); presence of white papillae on the bracts and sepals (when dense, capitulum appearing white); whether the female sepals are obtuse or acute, or thickened along the midline into a wing; and whether the petals are glandular or pilose. The status of these variants can only be determined by a study covering the whole range of the complex, so the name is applied here in a broad sense to cover all forms in the Flora area.

E. submersum Rendle differs from *E. setaceum* by its larger capitula (8 mm. wide) with bigger flowers (± 2 mm), and by its broader leaves (up to 0.8 mm. wide) which are often 3-nerved. It is known only from the types collected in Huila, Angola, and its status is very doubtful. It may be simply an unusually vigorous collection of *E. setaceum.*

2. **E. cinereum** *R. Br.*, Prodr. Fl. Nov. Holl.: 254 (1810); Meikle in F.W.T.A., ed. 2, 3: 63, fig. 337/21 (1968); Oberm. in F.S.A. 4(2): 11, fig. 2/2 (1985). Type: Australia, *R. Brown* (BM, holo.)

Small tufted annual. Leaves narrowly linear to acicular, setaceous-tipped, forming a neat dense cluster, 1–3 cm. long, 0.2–2 mm. wide, fenestrate. Scapes slender, straight, often numerous, 4–13 cm. high, 5-ribbed; sheaths ± equalling the leaves, subinflated, the mouth scarious, shortly obliquely slit. Capitulum globose to ovoid, 2–4 mm. diameter, grey- and straw-coloured or becoming darker, the bracts very loosely erect; involucral bracts as long as the capitulum, obovate-oblong, rounded becoming lacerate, ascending, pallid, scarious becoming tougher downwards; floral bracts lanceolate-oblong, thinly scarious, pale with a dark central band or sometimes blackish, glabrous or a few inner lightly keeled with fine sparse hairs on the keel; receptacle thinly pilose; flowers trimerous, pedicellate. Male flowers: sepals connate into a spathe with tridentate tip, dark grey, glabrous; petals included within the calyx, tiny; anthers white, 0.1 mm.

long. Female flowers much reduced and consisting mostly of the gynoecium: sepals 3, 2 or absent, filiform, caducous; petals absent, their position indicated by a node on the ovary stipe. Seeds 0.3 mm. long, light brown, glossy, faintly reticulate.

TANZANIA. Mwanza District: E. Uzinza [Uzinja], *Stuhlmann* 3552; Rufiji District: Mafia I., Kilindoni, 6 Aug. 1936, *Fitzgerald* 5213/1!; Songea District: Kwamponjore valley, 9.5 km. W. of Songea, 26 Apr. 1956, *Milne-Redhead & Taylor* 9834!

DISTR. **T** 1, 6, 8; scattered localities in West Africa from Senegal to Chad and in southern tropical Africa, but probably not native in Africa; India to China, Japan, southeast Asia and Australia; introduced to warm areas elsewhere as a weed of rice cultivation

HAB. Drying pool margins and seepages in lowland areas; sea-level–1000 m.

SYN. *E. sieboldianum* Steud., Syn. Pl. Glum. 2: 272 (1855), as '*sieboldtianum*'; Ruhland in E.P. 4, 30: 111, fig. 15A–G (1903). Type: Japan, *Siebold & Zuccarini* (LE, holo.)

E. amboense Schinz in Bull. Herb. Boiss. 4, App. 3: 35 (1896); N.E. Br. in F.T.A. 8: 258 (1901); Ruhland in E.P. 4, 30: 112 (1903); H. Hess in Ber. Schweiz. Bot. Ges. 65: 176, t. 9/3, fig. 1 on p. 160 (1955); Friedr.-Holzh. & Roessler in Prodr. Fl. SW.-Afr. 159: 2 (1967). Type: Namibia, Ovambo, Uashitenga near Olukonda, *Schinz* 859 (Z, holo., K, iso.)

E. heudelotii N.E. Br. in F.T.A. 8: 258 (1901). Types: Senegal/Mali [Senegambia], without precise locality, *Heudelot* 677 & 678 (K, syn.!)

E. stuhlmanni N.E. Br. in F.T.A. 8: 259 (1901). Type: Tanzania, Mwanza District, E. Uzinza [Uzinja], *Stuhlmann* 3552 (B, holo.)

NOTE. Distinctive on account of its tiny white anthers and reduced female flowers. The cushion of small acicular leaves is also characteristic.

3. **E. polhillii** *S.M. Phillips* in K.B. 51: 626, fig. 3E–H, 4A–B (1996). Type: Tanzania, Singida District, 30 km. from Issuna on the Singida–Manyoni road, *Greenway & Polhill* 11552 (K, holo.!)

Rosulate annual. Leaves linear-lanceolate to narrowly linear, 3–11 cm. long, 1.5–3 mm. wide, dark green, spongy, tip acuminate to filiform. Scapes up to 15, 9–15 cm. high, 5-ribbed, the ribs prominent and sharply angled; sheaths as long as the leaves, the limb acute, soon splitting. Capitulum hemispherical to slightly dome-shaped, 4–5 mm. diameter, blackish with pale involucral bracts, the narrow floral bracts and sepals intermingling; involucral bracts conspicuous, pallid, as wide as the capitulum and spreading in several series, narrowly lanceolate-oblong, 2.4–2.5 mm. long, scarious, acute, not reflexing; floral bracts dark grey, narrowly oblong, concave round the flowers, tip acute with denticulate margins; receptacle cylindrical, villous; flowers much reduced, glabrous. Male flowers 1.5 mm. long: sepals 2, free, blackish, oblanceolate-oblong, falcate, upper margins denticulate to incised, tip finely acute; petals absent; anthers black, 6 or often one or two aborted. Female flowers 2.0 mm. long: sepals 2, blackish, linear-falcate, lightly keeled, upper margins denticulate, tip finely acute; petals absent; ovary trilocular, raised on a slender stipe 0.5 mm. long, position of petals indicated by a node on the stipe. Seeds 0.4 mm. long, clear yellow, glossy, faintly reticulate and minutely papillose. Fig. 1/5–8; fig. 4/3.

TANZANIA. Singida District: 30 km. from Issuna on the Singida–Manyoni road, 13 Apr. 1964, *Greenway & Polhill* 11552! & 27 km. on Manyoni–Singida road, 3 July 1996, *Faden et al.* 96/542!; Dodoma District: 16 km. W. of Kazikazi, S. of Itigi–Tabora track, Chaya Lake, 2 July 1996, *Faden et al.* 96/509!

DISTR. **T** 5; not known elsewhere

HAB. Boggy places on sandy soil and shallow water of seepage and lake margins; 1250–1500 m.

NOTE. *E. polhillii* is unique among African *Eriocaulon* species in its combination of much reduced flowers lacking petals and black anthers. The only other species in Africa with similarly reduced flowers, with only two sepals and female flowers lacking petals, is *E. cinereum* R. Br., but this can easily be distinguished by its smaller habit with a dense cushion of acicular leaves, and by its white anthers. *E. polhillii* is probably related to *E. fuscum* S.M. Phillips from Zimbabwe, which has linear-falcate sepals and sometimes vestigial petals in the female flowers, together with a similar seed.

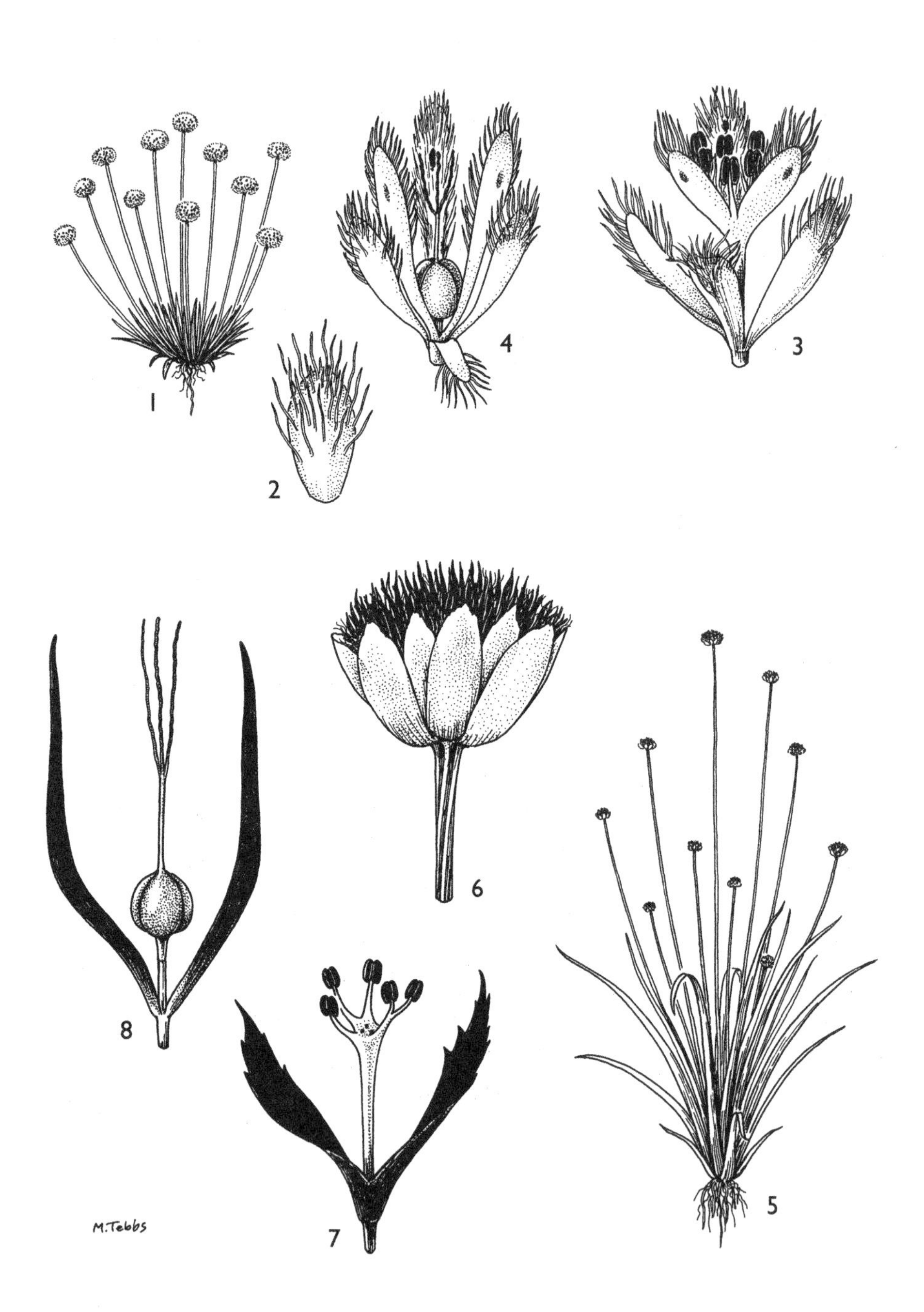

FIG. 1. *ERIOCAULON LANICEPS* — **1**, habit, × $^2/_3$; **2**, involucral bract, × 10; **3**, male flower, × 20; **4**, female flower, × 20. *ERIOCAULON POLHILLII* — **5**, habit, × $^2/_3$; **6**, capitulum, × 10; **7**, male flower, × 30; **8**, female flower, × 30. 1–4, from *Bullock* 3278; 5–8, from *Greenway & Polhill* 11552. Drawn by Margaret Tebbs.

4. **E. glandulosum** *Kimpouni* in Fragm. Fl. Geobot. 39: 336, fig. 10 (1994). Type: Zaire, Shaba, Dilolo, *Risopoulos* 1041 (BR, holo.!)

Small annual. Leaves many in a basal tuft, filiform with a setaceous tip, 1–2 cm. long, not expanded at the base. Scapes filiform, conspicuously papillose, up to ± 25, 4–10 cm. high, 3-ribbed; sheaths shorter than the leaves, not papillose. Capitulum depressed-globose with truncate base, 1–2 mm. diameter, the bracts thinly cartilaginous; involucral bracts straw-coloured, noticably paler than the floral bracts, obovate to suborbicular, 0.8–1.0 mm. long, rounded; floral bracts similar to the involucral but flushed grey or blackish, translucent, the upper margin crenulate; receptacle pilose; flowers dimerous, strongly laterally compressed, 0.8–0.9 mm. long, glabrous. Male flowers: sepals free, narrowly oblong, obtuse, denticulate on the upper margin; petals rudimentary; anthers 4, black. Female flowers: sepals strongly keeled, almost square in profile, overlapping and covering the petals and ovary, truncate, narrowly winged on the keel, a hyaline patch on the centre body, upper sepal-margin and wing-margin irregularly denticulate; petals unequal, narrowly oblanceolate, eglandular, the tips denticulate, sometimes emarginate, occasionally with one or two small marginal hairs. Mature seeds not seen.

TANZANIA. Songea District: Kwamponjore valley, 19 June 1956, *Milne-Redhead & Taylor* 10842!
DISTR. **T** 8; Zaire (Shaba)
HAB. Drying margins of temporary pools; 1000 m.

NOTE. Very similar in facies to *E. nigrocapitatum* but with even smaller capitula which are wider than long, and paler involucral bracts. The papillose scapes provide an easy spot character.

5. **E. mutatum** *N.E. Br.* in F.T.A. 8: 256 (1901); H. Hess in Ber. Schweiz. Bot. Ges. 65: 167, t. 9/1, fig. 9–10 on p. 160 (1955); S.M. Phillips in K.B. 51: 641–646, fig. 9A–B (1996). Types: Angola, Huila, between Lopollo and Monino, *Welwitsch* 2448 (BM, syn.!, K, isosyn.!) & 2449 & 2450 (both BM, syn.!)

Small rosulate annual. Leaves subulate to filiform, 1–3.5 cm. long, 0.2–1.0 mm. wide, tapering to a slender acuminate or setaceous tip. Scapes 10–± 50, filiform, 3–15 cm. high, 2–4-ribbed; sheaths shorter or longer than the leaves, inflated upwards and deeply slit, the limb obtuse to acute, sometimes splitting. Capitulum globose to slightly dome-shaped, 1.5–3(–3.5) mm. diameter, black, the bracts loose with the flowers visible among them; involucral bracts ± as wide as the capitulum, grey or infrequently light brown (lighter than rest of capitulum), scarious, 0.6–1.2 mm. long, obovate to subrotund with a rounded tip, spreading at maturity; floral bracts obovate to obovate-oblong, blackish, glabrous, obtuse to acute; receptacle glabrous or thinly hairy; flowers dimerous, blackish, 0.7–1.0 mm. long, completely glabrous. Male flowers: sepals free, narrowly oblanceolate-oblong, obliquely erose-truncate; petals unequal, very small and included within the calyx, eglandular, one 0.3 mm. long, emarginate, the other vestigial; anthers black. Female flowers: sepals very variable, concave and keeled, broad with gibbous overlapping margins and winged keel, the wing scarious, up to 0.6 mm. wide with a coarsely toothed margin and the upper sepal-margins also toothed, varying to narrowly falcate and unwinged or at most with a small tooth at the centre back, the tip acute to cuspidate; petals dimorphic, eglandular, the dorsal oblanceolate, 1.0 mm. long, entire to emarginate, the ventral shorter and narrower, linear-oblong, bidentate. Seeds ellipsoid, 0.3 mm. long, brown, smooth. Fig. 3/7.

var. **mutatum**; S.M. Phillips in K.B. 51: 643 (1996)

Leaves usually subulate, 0.4–1.2 mm. wide. Capitulum 2.5–3.0(–3.5) mm. diameter. Female sepals broad and enclosing the seed, winged on the keel, the wing up to 0.7 mm. wide, coarsely toothed.

TANZANIA. Songea District: Kwamponjore valley, 19 June 1956, *Milne-Redhead & Taylor* 10840! & N. of Songea, near Mshangano fishponds, 15 Apr. 1956, *Milne-Redhead & Taylor* 9920! & N. of Songea, by Lumecha bridge, 4 May 1956, *Milne-Redhead & Taylor* 9890!
DISTR. **T** 7, 8; as for species as a whole

SYN. *E. huillense* Rendle, Cat. Afr. Pl. Welw. 2: 95 (May 1899), *nom. illegit.*, *non* Engl. & Ruhland (Apr. 1899). Types as for *E. mutatum*
E. pseudomutatum Kimpouni in Fragm. Fl. Geobot. 39: 343, fig. 16 (1994). Type: Zaire, Shaba, near the source of the Luansoko, *Duvigneaud* 2954E (BRLU, holo.!)

var. **angustisepalum** (*H. Hess*) *S.M. Phillips* in K.B. 51: 644 (1996). Type: Angola, Huila, Guanhama, 45 km. S. of Cassinga, *Hess* 52/2004 (Z, holo., K, iso.!)

Leaves often filiform, 0.2–0.3 mm. wide. Capitulum 1.5–2.5 mm. wide. Female sepals narrow and exposing the seed, the keel unwinged or at most with a tooth up to 0.1 mm. wide.

TANZANIA. Tabora District: S. of Pozo Moyo, ± 8 km. from Kaliua, 22 June 1980, *Hooper & Townsend* 2102A!; Songea District: without precise locality, 27 June 1956, *Milne-Redhead & Taylor* 10914!
DISTR. **T** 4, 8; as for species as a whole

SYN. *E. angustisepalum* H. Hess in Ber. Schweiz. Bot. Ges. 65: 170, t. 9/6–7, fig. 7–8 on p. 160 (1955); Oberm. in F.S.A. 4(2): 11, fig. 2/3 (1985).

DISTR. (of species as a whole) **T** 4, 7, 8; Zaire (Shaba), Angola, Zambia, Malawi, Zimbabwe, South Africa (Transvaal, Natal)
HAB. (of species as a whole) Shallow margins and muddy edges of drying pools, and in seepage areas over rocks; 1000–1050 m.

NOTE. This delicate annual, although seldom collected, occurs over a wide area of southern tropical Africa. The slender habit with small black capitula is very similar to that of *E. abyssinicum*, but it can be immediately distinguished by its dimerous flowers.

Specimens with narrow unwinged female sepals look strikingly different from the more usual form with a broad toothed wing, but every gradation exists between the two extremes and mixed populations sometimes occur. A distinction at specific level is untenable, but forms with ± unwinged sepals often (though not always) have somewhat smaller capitula and filiform leaves and are distinguished here at varietal level. A third variety, var. *majus* S.M. Phillips, with a more vigorous habit, capitula 4–4.5 mm. wide and very broadly winged sepals, occurs in Zambia.

6. **E. nigrocapitatum** *Kimpouni* in Fragm. Fl. Geobot. 39: 340, fig. 13 (1994). Type: Zaire, Shaba, Kundelungu Plateau, between Mukumbi and Swambo, *Duvigneaud* 2951 (BRLU, holo.!)

Slender rosulate annual. Leaves numerous, setaceous, flaccid, 1.5–4 cm. long, scarcely broadened at the base. Scapes up to 50 or more, 5–12(–15) cm. high, 3–4-ribbed; sheaths shorter than the leaves, inflated upwards, deeply slit. Capitulum globose to ovoid, 1.5–2 mm. diameter, blackish, the bracts loosely imbricate with the petal-tips visible; involucral bracts as wide as the capitulum, concolorous with the floral bracts, blackish, scarious, broadly obovate, 0.8–1.1 mm. long, rounded; floral bracts similar to the involucral, suborbicular, broadly obtuse with erose upper margin; receptacle shortly pilose; flowers dimerous, flushed grey upwards. Male flower 0.8–1.0 mm. long: sepals free, oblong-falcate, lightly keeled, denticulate-acute; petals triangular, eglandular, the larger 0.3 mm.; anthers 4, black. Female flower 1.2 mm. long: sepals narrowly oblong-falcate with the petals and ovary exposed, narrowly winged on the keel or merely inconspicuously thickened, sometimes one or two short hyaline hairs at the top of the wing, acute; petals slightly unequal, narrowly oblanceolate-spathulate, emarginate or denticulate, longer than the sepals. Seed ellipsoid, 0.3 mm. long, brown, thickly white-reticulate. Fig. 4/2.

TANZANIA. Songea District: Kwamponjore valley, 19 June 1956, *Milne-Redhead & Taylor* 10841!
DISTR. **T** 8; Zaire (Shaba)
HAB. Boggy margins of drying pools; 1000 m.

NOTE. *E. nigrocapitatum* can be recognised by its small black capitula and fine leaves. It is very similar in floral morphology to *E. mutatum* var. *angustisepalum* but is distinguished most easily by its quite different, very obviously white-reticulate seed. Its seed will also readily distinguish it from other small, black-headed annuals.

7. **E. truncatum** *Mart.* in Wall., Pl. As. Rar. 3: 29 (1832). Type: India, Bihar, Munger [Monghir] Hills, *Hamilton*, Wall. Cat. no. 6076 (K-WA, holo.!)

Tufted annual. Leaves linear to linear-lanceolate, up to 6 cm. long, 1.5–3 mm. wide, opaque, subacute and apiculate. Scapes many, strict, stout, up to 15 cm. high, 5-ribbed; sheaths equalling the leaves, inflated, obliquely slit, the limb acuminate, often splitting and becoming bifid. Capitulum crateriform, ± 5 mm. diameter, glistening white to yellowish grey; involucral bracts as wide as the capitulum, 2.5–2.7 mm. long, oblong with a rounded to obtusely triangular tip, shiny, scarious, pallid, spreading at maturity; floral bracts obovate, rounded, similar to the involucral bracts but usually shorter, glabrous; receptacle subglabrous to pilose. Flowers with trimerous ovary, 1.5–1.8 mm. long, pallid throughout. Male flowers: sepals 2, grey, oblong-spathulate, conduplicate, tips obtuse with a few white papillae; petals 3, tiny, glandular and white-papillose; anthers black. Female flowers: sepals 2, resembling the male; petals and ovary usually raised on a stipe up to 0.4 mm. long; petals 3, narrowly oblanceolate, pilose on the inner face with spreading, septate, hyaline hairs, glandular and white-papillose at the tip. Seeds ellipsoid, 0.5 mm. long, pale yellow to brownish yellow, longitudinally ribbed.

TANZANIA. Rufiji District: Mafia I., Dawe Simba to Ndaagoni, 4 Oct. 1937, *Greenway* 5389!; Uzaramo District: 16 km. from Dar es Salaam towards Kisarawe [Kiserawe], 17 July 1971, *Batty* 1325! & 25 km. SW. of Dar es Salaam on road to Chanika, 23 Sept. 1977, *Wingfield* 4246!
DISTR. **T** 6; Mozambique, India, China and SE. Asia
HAB. Wet sandy or muddy pool and stream margins; near sea-level

SYN. *E. annuum* Milne-Redh. in Hook., Ic. Pl. 34, t. 3389 (1939). Type: Tanzania, Rufiji District, Mafia I., Dawe Simba to Ndaagoni, *Greenway* 5389 (K, holo.!)
E. ciliipetalum H. Hess in Ber. Schweiz. Bot. Ges. 65: 263, fig. 1 & 3 on p. 265 (1955). Type: Tanzania, Mafia I., Ngombeni, 12 July 1932, *Schlieben* (Z, holo.!)

NOTE. *E. truncatum* is a small annual widespread in tropical Asia. The few specimens known from the eastern coast of Africa have more densely hairy female petals than Asian plants, but are otherwise indistinguishable. The longitudinally elongated cells are extremely unusual in *Eriocaulon*. They are plentifully produced and, coupled with the pale shiny capitula, are the easiest diagnostic character for this species.

The closely related species *E. schlechteri* Ruhland from the coastal region around Natal has similar ribbed seeds, but is easily separated by its smaller darker capitula (3.5–4 mm. diameter) and silky-ciliate margins of the female petals.

8. **E. pictum** *Fritsch* in Bull. Herb. Boiss., sér. 2, 1: 1102 (1901). Type: Angola, Huila, *Dekindt* 703 (W, holo.)

Perennial forming clusters of leaf rosettes from a short rhizome clothed in roots and old leaves. Leaves lanceolate, 4–10 cm. long, 3–5 mm. wide, thick and spongy, yellow-green, tapering uniformly to an acute, hard, yellow-brown tip, woolly in the axils with long matted hairs. Scapes few, often solitary, 25–65 cm. high, 0.6–0.9 mm. in diameter, 6–8-ribbed; sheaths usually clearly longer than the leaf rosette, loose, the mouth scarious, obliquely slit with a rounded limb or becoming lacerate. Capitulum depressed-globose, 8–12 mm. diameter, grey becoming white; involucral bracts shorter than the capitulum width, ± 3 mm. long, yellow, brown or coppery, cartilaginous, broadly elliptic to oblong with obtuse or lacerate tips, somewhat reflexed at maturity; outer floral bracts resembling the involucral, narrowly oblong-cuneate, acuminate, grading into the grey inner bracts, these scarious with a thicker midline, white-hairy below the cuspidate tip; receptacle pilose, sometimes sparsely; flowers trimerous. Male flowers ± 5 mm. long: sepals narrowly oblong, the two laterals keeled, third flat, connate into a grey spathe with free, white-hairy, subacute tips; petals ligulate, exserted from the calyx, unequal with the longest up to 3.5 mm. long, white-villous at the tips, white-pilose with shorter hairs on the inner blade; anthers brownish black. Female flowers: sepals resembling the male calyx, connate into a grey spathe with free, white-hairy tips; petals raised on a short villous stipe, unequal, narrowly oblanceolate, the longest 3 mm. long, clawed and clearly exserted from the calyx, the other two 2 mm. long, white-hairy at the tips, conspicuously villous with spreading, hyaline, septate hairs ± 1 mm. long from the lower blade; ovary sessile. Seeds ellipsoid, brown, 0.75 mm. long.

TANZANIA. Iringa District: near Mufindi, 14 Nov. 1958, *Napper* 911!; Njombe District: Njombe–Kipengere road, 1.5 km. beyond Igosi, 26 Apr. 1970, *Wingfield* 593! & Lupembe road, Nov. 1928, *Haarer* 1577!

DISTR. **T** 7; Angola, Zambia, Malawi (Nyika Plateau) and Zimbabwe

HAB. Seasonal pools and swampy ground, often in peat bogs in association with *E. teusczii*; 1600–2200 m.

SYN. *E. amphibium* Rendle in J.L.S. 37: 475 (1906). Type: Zimbabwe, Matopo Hills, near American Mission, *Gibbs* 210 (BM, holo.!, K, iso.!)

NOTE. *E. pictum* is often confused with *E. teusczii*, as in general facies the two species are extremely similar and also frequently grow together in the same peaty habitats. However, they can easily be distinguished by the female flowers, as the spathate female calyx of *E. pictum* with a brush of woolly hairs protruding from the slit, is very characteristic. They can also be distinguished by the leaf-tip which is obviously rounded in *E. teusczii*, although some care is needed as the pointed, brown leaf-tips of *E. pictum* are easily broken off. The leaves in *E. pictum* are yellow-green rather than blue-green as in *E. teusczii*. The immature capitula are dark-coloured as the pointed tips of the floral bracts intermingle among the small white flowers, but as the petals elongate at maturity the darker bracts become obscured and the capitulum turns fluffy white.

A spathate female calyx is very unusual in African *Eriocaulon*. The only other species in the Flora area where the sepals are sometimes connate is *E. schimperi*, but this is a much more robust species from montane areas above 2000 m.

9. **E. teusczii** *Engl. & Ruhland* in E.J. 27: 77 (Apr. 1899). Type: Angola, Malange, *Rensch* in *Mechow* 231 (B, holo.!, Z, iso.)

Perennial from a short rhizome, often robust. Leaves many, rush-like, narrow, semi-cylindrical, up to 40 cm. long, 1–3 mm. wide, usually crowded in an erect tuft, subulate when short, spongy, the tip obtuse with a pore on the upper surface, woolly in the axils. Scapes 1–6(–15), 25–65 cm. high, 6–8-ribbed; sheaths equalling or longer than the leaves, the mouth shortly slit, limb rounded or becoming split. Capitulum hemispherical becoming globose with intruded base at maturity, white, 8–12 mm. diameter, the pointed tips of the floral bracts usually visible among the exserted white-hairy petals; involucral bracts shorter than the capitulum width, in several series, spreading, firm, 2.3–3.5 mm. long, straw-coloured or light brown and often flushed grey or black with a central paler stripe, oblong to ovate, outer rounded, the inner acute; floral bracts up to 3.7 × 1.2–1.5 mm., cuneate, white-villous on the widest part, firm and straw-coloured below, the tip chartaceous, dark grey, brown or infrequently pale, often abruptly long-cuspidate but sometimes merely acute or apiculate; receptacle pilose; flowers trimerous, 3.0–4.5 mm. long. Male flowers: sepals free, ± equal, laterals narrowly oblong-navicular, 1.7–2.5 mm. long, white or with greyish or brownish tips, white-villous towards the obliquely truncate tip, median similar but flat; petals unequal, glandular, white-villous over the whole inner blade, two small, the third narrowly oblong, up to 3 mm. long, exserted from the capitulum; anthers black. Female flowers: sepals as in the male flowers, 1.5–2.8 mm. long; petals and ovary both stipitate; petals unequal, glandular, white-villous over the inner blade, narrowly oblanceolate-oblong, somewhat narrowed to the base but not clawed, two 1.5–2.5 × 0.5–0.6 mm., the third ± $^1/_3$ longer, exserted from the capitulum. Seeds subrotund, 0.5–0.6 mm. long, reddish brown with white papillae.

TANZANIA. Iringa District: above Lake Ngwazi, 14 Sept. 1971, *Perdue & Kibuwa* 11411!; Songea District: Kwamponjore valley, 14 Mar. 1956, *Milne-Redhead & Taylor* 9163!; Ufipa District: Sumbawanga, Chapota swamp, 6 Mar. 1957, *Richards* 8512!

DISTR. **T** 4, 7, 8; N. Nigeria, Zaire (Shaba), Angola, Zambia, Malawi and Zimbabwe

HAB. Common and sometimes locally dominant in wet swamps among coarse grass and in ditches, sometimes in shallow standing water; 950–1800 m.

SYN. *E. huillense* Engl. & Ruhland in E.J. 27: 78 (Apr. 1899), *non* Rendle (May 1899). Type: Angola, Huila, May 1895, *Antunes* (B, holo.!)

E. lacteum Rendle, Cat. Afr. Pl. Welw. 2: 99 (May 1899). Types: Angola, Huila, Lopollo, *Welwitsch* 2452 & Morro de Lopollo, *Welwitsch* 2452b & Humpata District, slopes of Serra de Oiahoia, *Welwitsch* 2453 (all BM, syn.!, LISU, isosyn.)

E. intrusum Meikle in K.B. 22: 141 (1968) & in F.W.T.A., ed. 2, 3: 62, fig. 337/2 (1968). Type: Nigeria, Naraguta, *Lely* 283 (K, holo.!)

NOTE. *E. teusczii* is a very variable species vegetatively, especially in leaf length. Populations also differ in the colour of the involucral and floral bracts and sepal-tips, inconstant characters on which new species were based by earlier workers. The floral bracts are thinly to densely white-hairy and usually long-cuspidate. When dark-coloured the long tips are conspicuous among the white flowers in the capitulum. The capitula are often dark at first, but become fluffy-white as the petals are exserted. The combination of large fluffy-white capitula, together with unequal petals, free truncate-tipped male sepals, and obtuse leaves tipped with a pore, distinguishes *E. teusczii* from the other robust species.

E. teusczii is often confused with *E. pictum* Fritsch, with which it sometimes grows (see note under *E. pictum*).

10. **E. iringense** *S.M. Phillips* in K.B. 51: 337, fig. 1, 2E–F (1996). Type: Tanzania, Iringa District, Mafinga [Sao Hill], *Polhill & Paulo* 1720 (K, holo.!)

Robust rosulate perennial from a stout rhizome. Leaves broadly linear, tapering, 20–35 cm. long, 13–18 mm. wide, light green, thick, slightly glossy, the tip rounded. Scapes few, stout, 35–60 cm. high, 10-ribbed; sheaths almost equalling the leaves. Capitulum subglobose, 11–13 mm. diameter, greyish white; involucral bracts shorter than the capitulum width, oblong, yellow-brown, coriaceous, neatly imbricate in 3 series, subacute; floral bracts spathulate-angulate, firm, flushed blackish with paler, abruptly acute tip, white-pilose and concave at the widest part; receptacle villous; flowers trimerous, 3.0–3.5 mm. long. Male flowers: sepals oblanceolate, free or almost so, 2.2 × 1.0 mm., blackish upwards, lightly keeled, white-pilose at the tips; petals unequal, exserted from the calyx, glandular and villous, the hairs white above the gland and hyaline below, the longest petal 2.5 × 0.8 mm., the other two 1.5 × 0.5 mm.; anthers dark brown. Female flowers: sepals equal, blackish, oblanceolate-oblong, boat-shaped, 2 × 1 mm., white-pilose at the tip, villous around the keel with longer hyaline hairs; petals and ovary sessile; petals subequal, oblanceolate-oblong, densely villous with white hairs above the gland and longer hyaline hairs below, two 2.3 × 0.7 mm., the third slightly larger. Seeds subglobose, 0.75 mm. long, brown with many small hair-like projections. Fig. 2/1–7.

TANZANIA. Iringa District: Mafinga [Sao Hill], 12 Mar. 1962, *Polhill & Paulo* 1720! & Mufindi, Ngwazi Estate, near Ngwazi house, 3 Sept. 1971, *Perdue & Kibuwa* 11384!

DISTR. **T** 7; not known elsewhere

HAB. Marshy streamsides; ± 1900 m.

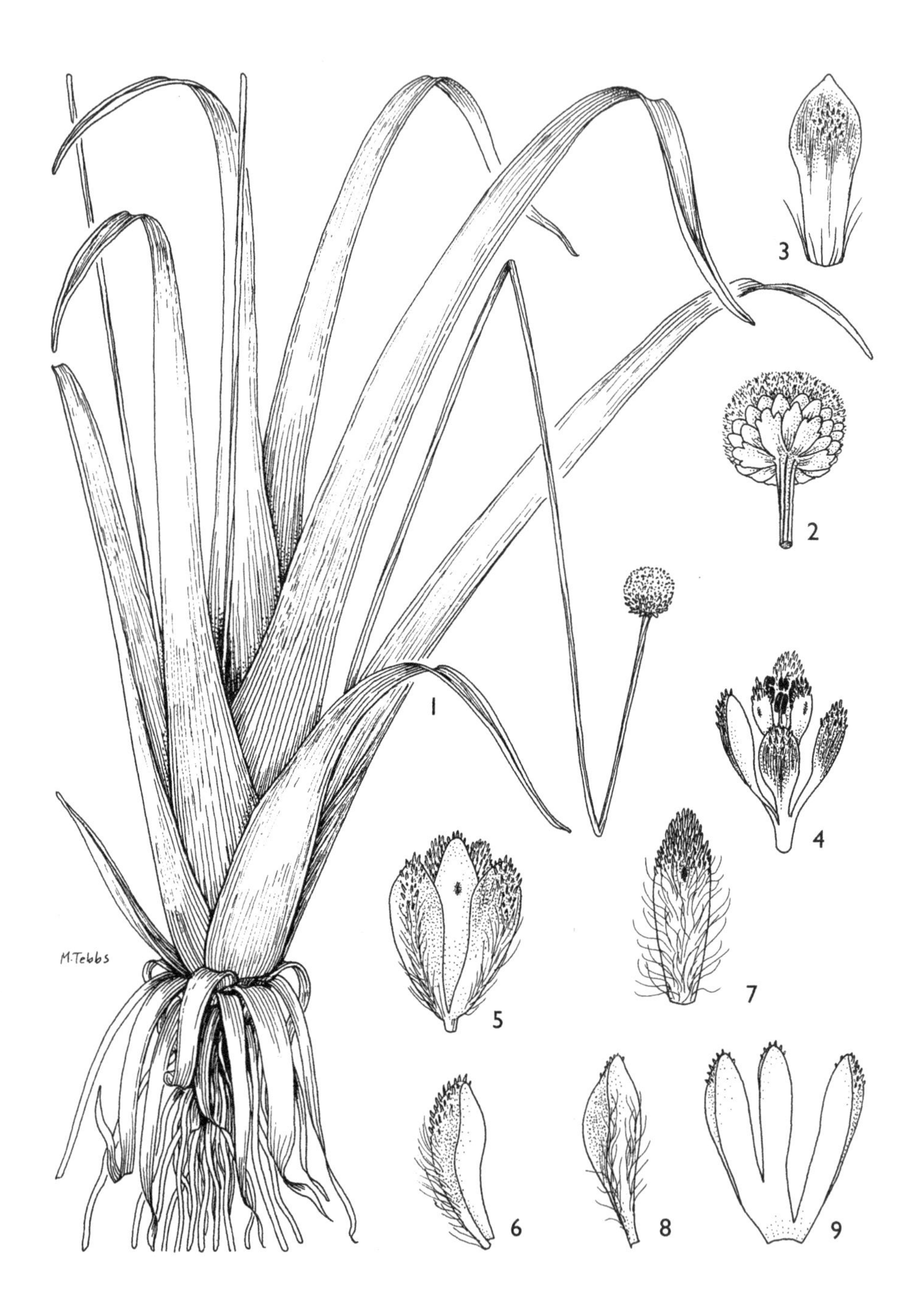

FIG. 2. *ERIOCAULON IRINGENSE* — **1**, habit, × $^2/_3$; **2**, capitulum, × 2; **3**, floral bract, × 10; **4**, male flower, × 10; **5**, female flower, × 10; **6**, female sepal, × 10; **7**, female petal, × 10. *ERIOCAULON MESANTHEMOIDES* — **8**, female sepal, × 10. *ERIOCAULON SCHIMPERI* — **9**, female sepals, × 10. 1–7, from *Polhill & Paulo* 1720; 8, from *Bruce* 723; 9, from *Scott* 153. Drawn by Margaret Tebbs.

NOTE. This species has the robust habit of *E. schimperi* and *E. mesanthemoides*, and is easily confused with both of these. It differs by its regular coriaceous involucral bracts, and especially by the villous receptacle and flower parts, which are immediately obvious when the capitulum is opened up.

11. **E. schimperi** *Ruhland* in E.J. 27: 80 (1899); N.E. Br. in F.T.A. 8: 243 (1901); Champl. in Fl. Rwanda 4: 147, fig. 53 (1988); Blundell, Wild Fl. E. Afr., reprint: 415, t. 138 (1992); U.K.W.F., ed. 2: 303 (1994); S.M. Phillips in K.B. 51: 334, fig. 2A–B (1996); Lye in Fl. Ethiopia 6: 379, fig. 209.1/1–4 (1997). Type: Ethiopia, Jan Meda [Dschan-Meda], *Schimper* 1217 (B, holo., K, iso.!)

Robust perennial from a stout erect rhizome. Leaves clustered at the rhizome tip, broadly linear, 6–25(–40) cm. long, 4–14(–18) mm. wide, light green, thick, smooth, the tip broadly rounded. Scapes 2–5, stout, straight, up to 35(–60) cm. high, 8–10-ribbed; sheaths shorter than the leaves, subinflated at the mouth and split into 2–3 papery lobes. Capitulum hemispherical, 9–14(–18) mm. wide, greyish white; involucral bracts in 2–3 series slightly narrower than the capitulum width, blackish, firmly scarious, elliptic-oblong, acute, glabrous; floral bracts oblong-cuneate, grey, densely white-pilose upwards; receptacle glabrous; flowers 3–3.5(–4.5) mm. long. Male flowers: sepals oblong-obovate, connate into a spathe with free tips, lightly keeled, black, white-pilose on the upper back and margins, broadly acute; petals glandular and white-villous above the middle, unequal, the longest 1.5–2.5 mm. long, its tip exserted from the calyx; anthers black. Female flowers: sepals blackish, subequal, variable, oblong to obovate-oblong, spongy along the midline, white-pilose towards the tips, free or frequently connate at the base, sometimes unequally connate to the middle or above and resembling the male calyx; petals and ovary subsessile; petals white-pilose above the middle, slightly unequal. Seeds ellipsoid to subrotund, 0.7–1.0 mm. long, brown. Fig. 2/9.

UGANDA. Kigezi District: Behungi swamp, 1 Dec. 1930, *B.D. Burtt* 2922! & Muchoya Fen Bamboo Reserve, 5 Jan. 1962, *Morrison* 18! & Mgahinga [Gahinga]-Muhavura saddle, 24 Apr. 1970, *Lye* 5286!

KENYA. Elgeyo District: Cherangani Hills, 3 km. from the turn off of the Kapsangar track from the Kaibibich track, 17 Apr. 1975, *Friis & Hansen* 2533!; Ravine District: Timboroa, highland near sawmill, 14 June 1953, *G.R. Williams* 562!; N. Kavirondo District: Elgon, SE. side, track to Laboot, 16 Mar. 1977, *Hooper & Townsend* 1392!

TANZANIA. Mbeya District: Mbeya Peak public land, 26 Oct. 1958, *Gaetan* 89!; Iringa District: Dabaga Highlands, Kibengu, 20 Feb. 1962, *Polhill & Paulo* 1540!; Njombe District: Poroto Mts., Kitulo Plateau, Ndumbi valley, 24 Mar. 1991, *Bidgood, Congdon & Vollesen* 2109!

DISTR. **U** 2; **K** 3, 5; **T** 7; Sudan (Imatongs), Ethiopia, Zaire (E. mountains), Rwanda, Burundi and Malawi (Nyika Plateau)

HAB. Wet afroalpine grassland, sometimes with *Dendrosenecio*, marshy streamsides, and in peat bogs with sphagnum and sedges; 2000–3500 m.

SYN. *E. schimperi* Ruhland var. *gigas* Moldenke in Phytologia 2: 364 (1947). Type: Kenya, Elgeyo District, Marakwet Hills, Moyben R., *Dale* in *F.D.* 3397 (BR, holo., K, iso.!)

NOTE. A robust species from the mountains, characterised by a large rosette of thick, blunt leaves and a few stout scapes bearing large capitula. The different populations show small variations, especially in the texture and colour of the involucral bracts, the degree of hairiness of the floral bracts and flowers leading to differences in capitulum colour, the form of the female calyx, and the size of the petal glands. The female sepals are frequently spathately connate to above the middle in populations from mountains bordering the western Rift Valley from Burundi northwards to the Imatong mountains of southern Sudan (S.M. Phillips in K.B. 51: 335 (1996)).

12. **E. mesanthemoides** *Ruhland* in E.J. 27: 79 (1899); N.E. Br. in F.T.A. 8: 244 (1901); S.M. Phillips in K.B. 51: 336 (1996). Type: Tanzania, Morogoro District, Uluguru Mts., Lukwangule Plateau, Ukami, 6 Nov. 1894, *Stuhlmann* 9143 (B, holo.!)

Robust tufted perennial from a short rhizome. Leaves narrowly lanceolate, 7–30 cm. long, 6–10 mm. wide, firm, gradually tapering to a narrowly obtuse tip. Scapes 1–5, stout, 10–60 cm. high, 8–9-ribbed; sheaths usually shorter than the leaves, loose, the limb subacute. Capitulum depressed-globose, 12–19 mm. diameter, white speckled with darker bracts; involucral bracts large and scarious, as wide as the capitulum, pallid, flushed grey or blackish, ovate to ovate-oblong, 4.5–6.0 mm. long, acute, finally reflexing; floral bracts narrowly oblong, flushed grey, concave, densely white-villous above the middle, long-acuminate; receptacle thinly hairy; flowers trimerous, 3.5–4.5 mm. long. Male flowers: sepals grey, unequally connate below or up to the middle, the two lateral lightly keeled, the third concave, white-pilose across the tips; petals ligular, exserted from the calyx, glandular and white-villous, dimorphic, the longest 2.1–3.5 mm. long; anthers black. Female flowers: sepals oblong, grey, deeply concave, thickened and spongy along the midline and sometimes expanded into a narrow wing, villous along the lower margins inside, white-pilose at the tip; petals and ovary sessile; petals oblong-spathulate, glandular and white-villous upwards, the mid-petal longer. Seeds ellipsoid to broadly ellipsoid, 0.8–1.0 mm. long, mid-brown or reddish brown. Fig. 2/8.

KENYA. Nyeri District: Aberdare National Park, E. side of plateau, 9 Apr. 1975, *Hepper & Field* 4943! & near W. part of Nyeri track, 16 July 1948, *Hedberg* 1602! & Karimu R. near the Gura Falls, 14 Mar. 1964, *Verdcourt* 3997!

TANZANIA. Morogoro District: Uluguru Mts., Lukwangule Plateau, 19 Sept. 1970, *Thulin & Mhoro* 1051! & 2 Jan. 1975, *Polhill & Wingfield* 4667! & 27 Jan. 1976, *Cribb & Grey-Wilson* 10469.!

DISTR. **K** 3, 4; **T** 6; Malawi (Nyika Plateau)

HAB. Swampy ground and wet streamsides in montane grassland or moorland or in forest clearings; 2400–3100 m.

SYN. *E. friesiorum* Bullock in K.B. 1932: 507 (1932). Type: Kenya, Aberdare Mts., *R.E. & T.C.E. Fries* 2402 (K, holo.!)

NOTE. *E. mesanthemoides* appears to be almost restricted to the Ulugurus in Tanzania and the Aberdares in Kenya, but is so similar in facies and habitat requirements to the commoner *E. schimperi* that it may have been overlooked elsewhere in East Africa. The pointed involucral bracts are larger and thinner than in *E. schimperi*, radiating across the full width of the capitulum base and the leaves are more tapering, especially in larger specimens. The female sepals of *E. schimperi* are never villous inside.

13. **E. volkensii** *Engl.*, P.O.A. C: 133 (1895); N.E. Br. in F.T.A. 8: 238 (1901); A.V.P.: 60, 263 (1901); U.K.W.F., ed. 2: 303 (1994); S.M. Phillips in K.B. 51: 339, fig. 2C–D (1996). Type: Tanzania, Kilimanjaro, Kibo, N. slope, *Volkens* 2032 (B, holo., BR, K!, iso.)

Low perennial, forming clusters of small rosettes from a branching rhizome. Leaves lanceolate, tapering from broad base to narrow obtuse tip, 3–7 cm. long, 5–8 mm. wide, firm, outwardly curving. Scapes up to ± 6, very short, the capitula remaining in the leaf rosette and sometimes subsessile, 3–4-ribbed; sheaths open nearly to the base, embedded among the woolly leaf-bases. Capitulum dark grey with light straw-coloured involucral bracts, 5–9 mm. diameter; involucral bracts forming a campanulate cup around the flowers, coriaceous, 2.2–3.2 mm. long, suborbicular, rounded across the thinner tips, spreading at maturity; floral bracts narrowly obovate, flushed grey, thinly white-pilose, subacute; receptacle glabrous; flowers trimerous, 2.6–2.8 mm. long. Male flowers: sepals dark grey, obovate-oblong, unequally connate below, white-pilose on the upper back; petals subequal, glandular and white-villous; anthers black. Female flowers: sepals blackish, oblong to obovate-oblong, concave and indistinctly keeled, white-pilose at the subacute tips; petals subequal, narrowly oblong-spathulate, glandular and white-pilose at the tips, pilose with hyaline hairs on the inner face. Seeds broadly ellipsoid, 0.7 mm. long, dark reddish brown.

KENYA. Elgeyo District: Cherangani Hills, Embobut R., near Kameligon, 26 Aug. 1969, *Mabberly & McCall* 229!; Aberdare Mts., along Magura R., downstream from Queen's Falls, 22 June 1974, *R.B. & A.J. Faden* 74/846!; Mt. Kenya, Naromoru side, 11 Dec. 1957, *Verdcourt* 2001!
TANZANIA. Kilimanjaro, E. of Jonsell Point, 13 Jan. 1970, *Lye & Katende* 4841!
DISTR. **K** 3, 4; **T** 2; not known elsewhere
HAB. Montane grassland and moor, in boggy places and wet streamsides; 3000–3900 m.

NOTE. Small plants with very similar facies from above 3000 m. in the Ethiopian highlands are not this species, but dwarfed specimens of *E. schimperi*. They can be distinguished by their different involucral bracts and seed-coat patterning.

14. **E. laniceps** *S.M. Phillips* in K.B. 51: 625, fig. 1A–D, 2A–B (1996). Type: Tanzania, Kigoma District, 60 km. S. of Uvinza [Uvinsa], *Bullock* 3278 (K, holo.!)

Small rosulate annual. Leaves subulate, 1.2–1.8 cm. long, ± 1 mm. wide, fenestrate, acute. Scapes up to ± 20, slender, 3–5 cm. high, 4-ribbed; sheaths equalling the leaves, inflated, the limb subacute becoming denticulate. Capitulum depressed-globose, 4–6 mm. diameter, dirty white, fluffy with woolly hairs; involucral bracts as wide as the capitulum, pallid or greyish, scarious, obovate, 2 mm. long, villous with spreading, hyaline, tubercle-based, septate, twisted hairs, a few stouter opaque white hairs at the acute tip, not reflexing at maturity; floral bracts resembling the involucral but narrower, oblanceolate, villous with long hyaline hairs on the back and with white stout hairs at the tip; receptacle thinly pilose; flowers trimerous, 2.0–2.3 mm. long, pallid, subsessile. Male flowers: calyx variable, the sepals almost free, narrowly oblong, white-villous above the middle, the two lateral keeled and connate with the flat median sepal only at the base, or variably connate with the median sepal sometimes almost to the tips; petals subequal, the tips exserted from the calyx, glandular and white-villous. Female flowers: sepals narrowly oblong, lightly keeled, white-villous towards the tips; petals subequal, narrowly oblanceolate-oblong, glandular and white-villous on the inner face with hairs 0.3–0.4 mm. long. Seeds ellipsoid, 0.4 mm. long, brown, longitudinally white striate-punctate. Fig. 1/1–4.

TANZANIA. Kigoma District: 60 km. S. of Uvinza [Uvinsa], 31 Aug. 1950, *Bullock* 3278! & 56 km. on Uvinza–Mpanda road, 21 May 1997, *Bidgood et al.* 4110!; Mpanda District: 3 km. on Ugalla R. road from Mpanda–Uvinza road, 19 May 1997, *Bidgood et al.* 4040!
DISTR. **T** 4; not known elsewhere
HAB. Seepage areas and peaty bog overlying rock; 1100–1700 m.

NOTE. The woolly capitula are very distinctive and it is unlikely to be confused with any other East African species. The fluffy hyaline hairs from the involucral bracts surround the young capitulum and are still very obvious at maturity around the top of the scape.

In floral and vegetative morphology *E. laniceps* closely resembles *E. albocapitatum* Kimpouni from Zaire and Cameroon, but this species has glabrous involucral bracts. A single collection from **T** 4 with a glabrous involucre (*Bidgood et al.* 4136) has a different seed patterning from *E. albocapitatum* from Cameroon. Unfortunately the type of *E. albocapitatum* from Zaire lacks seed.

15. **E. elegantulum** *Engl.*, P.O.A. C: 133 (1895); N.E. Br. in F.T.A. 8: 254 (1901); Meikle in F.W.T.A., ed. 2, 3: 63 (1968). Type: Tanzania, Tanga District, Duga, *Holst* 3181 (B, holo., K, iso.!)

Slender tufted annual. Leaves linear, 2–7.5 cm. long, 1.5–6 mm. wide, thin, acute or less often acuminate. Scapes varying from few to ± 50, often slightly flexuous, 7–25 cm. high, 4-ribbed, the ribs forming narrow wings; sheaths ± as long as the leaves, the limb with a delicate hyaline acute tip, soon splitting. Capitulum globose, 3–5 mm. diameter, greyish white, the pale female petals showing among the white-hairy floral bracts; involucral bracts shorter than the capitulum width, 4–5 spaced in one circlet (best seen in very young, still hemispherical capitula), ovate-oblong, obtuse, thin and hyaline, crumpled and obscured by the expanding flowers and apparently absent at maturity; floral bracts narrowly oblong-cuneate, black, scarious, concave but not keeled, densely white-pilose on the upper blade, acute; receptacle pilose, often thinly. Flowers 3-merous, 1.0–1.2 mm. long, mainly female. Male flowers: calyx spathate, black, the free sepal-tips obtuse, white-pilose; petals small and included within the calyx, white-pilose, glands very small; anthers black. Female flowers: sepals black, narrowly oblong, one slightly to definitely smaller, concave, lightly keeled and narrowly winged above the middle, white-hairy towards the acute tip; petals and ovary subsessile; petals narrowly oblanceolate-oblong, subequal, pallid, white-hairy around the tips, glands small, often absent on the median petal. Seeds broadly ellipsoid, 0.3 mm. long, yellowish brown, translucent, white-reticulate. Fig. 4/4.

KENYA. Kwale District: Shimba Hills, Lango ya Mwagandi [Longo Mwagandi], 17 Mar. 1968, *Magogo & Glover* 355! & near Lunguma, 20 Aug. 1994, *Luke & Gray* 4056!; Kilifi District: 24 km. S. of Malindi, Mida, 3 Dec. 1961, *Polhill & Paulo* 901!

TANZANIA. Shinyanga, near aerodrome, 6 May 1931, *B.D. Burtt* 2437!; Tanga District: flats by Lwengera R., 6.5 km. ENE. of Korogwe, 27 June 1953, *Drummond & Hemsley* 3063; Masasi District: near Masasi, Chironga Hill, 8 Mar. 1991, *Bidgood, Abdallah & Vollesen* 1842!; Zanzibar, Maji Mekundu, 24 Jan. 1929, *Greenway* 1164!; Pemba, *Vaughan* 464!

DISTR. **K** 7; **T** 1, 3, 6–8; **Z**; **P**; S. Sudan, Malawi, Mozambique and Zimbabwe, also Ghana and S. Nigeria (?introduced)

HAB. Seasonal ponds, ditches, drainage lines and swampy grass areas, usually on sand, occasionally in rice fields; sea-level–1400 m.

NOTE. *E. elegantulum* can be difficult to distinguish from subsp. *hanningtonii* of *E. transvaalicum*, another annual with a similar facies of thin linear leaves and greyish capitula on slender scapes. *E. transvaalicum* differs by its 5-ribbed unwinged scapes, involucral bracts in more than one series which remain visible in the mature capitulum, and larger seeds (0.5 mm.).

16. **E. bongense** *Engl. & Ruhland* in E.J. 27: 75 (1899); N.E. Br. in F.T.A. 8: 246 (1901); Ruhland in E.P. 4, 30: 100 (1903); Meikle in F.W.T.A., ed. 2, 3: 63, fig. 338/18 (1968).Type: Sudan, Bongoland, Bulu stream, near Sabbi, *Schweinfurth* 2722 (B, holo., K, iso.!)

Rosulate herb, fairly robust, probably annual. Leaves ensiform, 3–6 cm. long, 4–7 mm. wide, spongy, subacute. Scapes 1–16, flexuous, 10–35 cm. high, 5–7-ribbed; sheaths exceeding the leaves, slightly inflated, the mouth deeply obliquely slit with an acute limb. Capitulum globose to conical, 5–8 mm. diameter, shaggy, shining pale silvery-grey with yellowish involucral bracts; involucral bracts forming a crateriform base to the capitulum, scarious with coriaceous lower part, obovate, 3.0–3.5 mm. long, broadly rounded with small central cusp, reflexing at maturity; floral bracts oblanceolate to cuneate-oblong, scarious, glabrous, pallid flushed grey, incurving, the tip extended into a long slender cusp; receptacle pilose; flowers trimerous, 2.0–2.5 mm. long, pedicellate. Male flowers: sepals free, linear, subequal or one slightly smaller, pallid to grey, glabrous; petals tiny, glabrous or white-pilose, eglandular; anthers dark. Female flowers: sepals narrowly linear, slightly geniculate, subequal or one noticeably smaller, pallid or grey, glabrous, acuminate; petals and ovary raised on a stipe; petals linear, white, subequal or the odd petal up to a third larger, eglandular, white-pilose at the tips. Seeds ellipsoid, 0.35 mm. long, brown, reticulate with white papillae outlining the reticulations. Fig. 3/8; fig. 4/6.

TANZANIA. Ufipa District: 14 km. on Matai–Nkowe road, 22 June 1996, *Faden et al.* 96/329!; Ulanga District: Selous Game Reserve, ± 8 km. S. of Ruaha camp, 20 Oct. 1975, *Vollesen* in *M.R.C.* 2896A!; Songea District: 9.5 km. SW. of Songea, valley near Mtanda, 24 June 1956, *Milne-Redhead & Taylor* 10893!

DISTR. **T** 4, 6–8; West Africa east to Sudan, also in Zambia

HAB. Marshy grassland, wet flushes and in rice fields; 250–1700 m.

NOTE. A distinctive species on account of the rather shaggy-looking capitula, the very narrow sepals and petals intermingling with the slender cusps of the floral bracts. The main centre of distribution is in West Africa, where the capitula are usually whitish yellow or light silvery-grey. The few collections from Tanzania and Zambia have darker grey capitula.

17. **E. abyssinicum** *Hochst.* in Flora 28: 341 (1845); N.E. Br. in Fl. Cap. 7: 53 (1897) & in F.T.A. 8: 257 (1901); Ruhland in E.P. 4, 30: 282 (1903); H. Hess in Ber. Schweiz. Bot. Ges. 65: 165, t. 9/8, fig. 2–3 on p. 160 (1955); Meikle in F.W.T.A., ed. 2, 3: 63, fig. 338/19 (1968); U.K.W.F., ed. 2: 303 (1994); Lye in Fl. Ethiopia 6: 379, fig. 209.1/8–9 (1997). Type: Ethiopia, Shire, *Schimper* 1944 (TUB, holo., K!, G, iso.)

Small rosulate annual. Leaves few, light green, subulate, relatively long, up to 5 cm. long, 0.8–1.7 mm. wide, exceeding the sheaths and often intermingling with the lower capitula, fenestrate, tapering to a fine acuminate tip. Scapes very slender, several to numerous and radiating to form a dome-shaped mound, 2–12 cm. high, 0.2–0.3 mm. in diameter, 3–4-ribbed; sheaths shorter than the leaves, the mouth oblique with entire subacute limb. Capitulum globose to ovoid, 2–3 mm. diameter, grey to blackish with paler involucral bracts, loose and untidy with prominent stigmas; involucral bracts as wide as the capitulum, pallid, scarious, 1.2–1.4 mm. long, elliptic to narrowly ovate, subacute, not reflexing; floral bracts oblanceolate-oblong, concave, subacute to shortly acuminate, scarious, flushed grey with the upper usually darker; receptacle glabrous to villous; flowers trimerous, ± 1 mm. long. Male flowers: calyx spathate, grey, sepals connate below the middle, the free tips triangular, glabrous, acute; petals rudimentary, glabrous; anthers black. Female flowers: sepals variable, subequal, usually flushed grey, narrowly lanceolate and slightly concave, varying to broader and deeply concave but not keeled, acute or acuminate, the margins glabrous, with a few short hairs, or occasionally conspicuously ciliate, the hairs 0.15–0.3 mm. long; petals subequal, linear, glabrous, vestigial glands usually present at least on the two smaller, sometimes emarginate, especially the larger. Seeds ellipsoid to broadly ellipsoid, 0.35 mm. long, brown, glossy, ± smooth with a faint reticulation. Fig. 3/4–6; fig. 4/1.

UGANDA. Mbale District: Elgon, Kapchorwa, 7 Sept. 1954, *Lind* 247! & N. slopes of Elgon, 1.5 km. SW. of Kapchorwa resthouse, 12 Oct. 1952, *G.H.S. Wood* 446! & Bugisu [Bugishu], Sipi, 31 Aug. 1932, *A.S. Thomas* 452!

KENYA. Machakos District: S. end of Mua Hills, 2 Feb. 1969, *Napper & Faden* 1862!; Nairobi District: Nairobi, golf range between Wilson airport and army barracks, 12 Feb. 1978, *M.G. Gilbert* 4976! & Langata Forest where Langata road crosses Mokoyeti R., 3 June 1951, *Rayner* 475!

TANZANIA. Moshi District: Mpololo, Aug. 1928, *Haarer* 1540!; Iringa, N. of township, 14 July 1956, *Milne-Redhead & Taylor* 11148!; Songea District: near Lumecha Bridge, 4 May 1956, *Milne-Redhead & Taylor* 9996!

DISTR. **U** 3; **K** 1, 3, 4; **T** 2, 4, 7, 8; tropical Africa from Ethiopia to Namibia and South Africa, also in N. Nigeria

HAB. On shallow muddy soil overlying rock outcrops, usually with water flowing over, in the mist zone of waterfalls, and in seasonal wet flushes and ponds in grassland, locally abundant in association with other small ephemerals including *E. cinereum* and *E. mutatum*; 950–2750 m.

SYN. *E. gilgianum* Ruhland in E.J. 27: 84 (Apr. 1899). Type: Angola, Huila, *Antunes* 168a (B, holo.!)

E. ciliisepalum Rendle, Cat. Afr. Pl. Welw. 2: 98 (May 1899). Types: Angola, Huila, near Lopollo, *Welwitsch* 2445 (BM, syn.) & Morro de Lopollo, *Welwitsch* 2445b (BM, syn.!, K, isosyn.!)

E. subulatum N.E. Br. in F.T.A. 8: 255 (1901). Type: Zimbabwe, Zambesi R., island at Victoria Falls, 1860, *Kirk* (K, holo.!)

NOTE. *E. abyssinicum* is a widely distributed and variable small pioneer species, recognised by its numerous slender scapes topped by small dark capitula with paler involucral bracts, and surrounded by a rosette of rather long, narrow, finely tapering leaves. The main source of variation lies in the degree of hairiness of the receptacle and female sepal-margins. It has

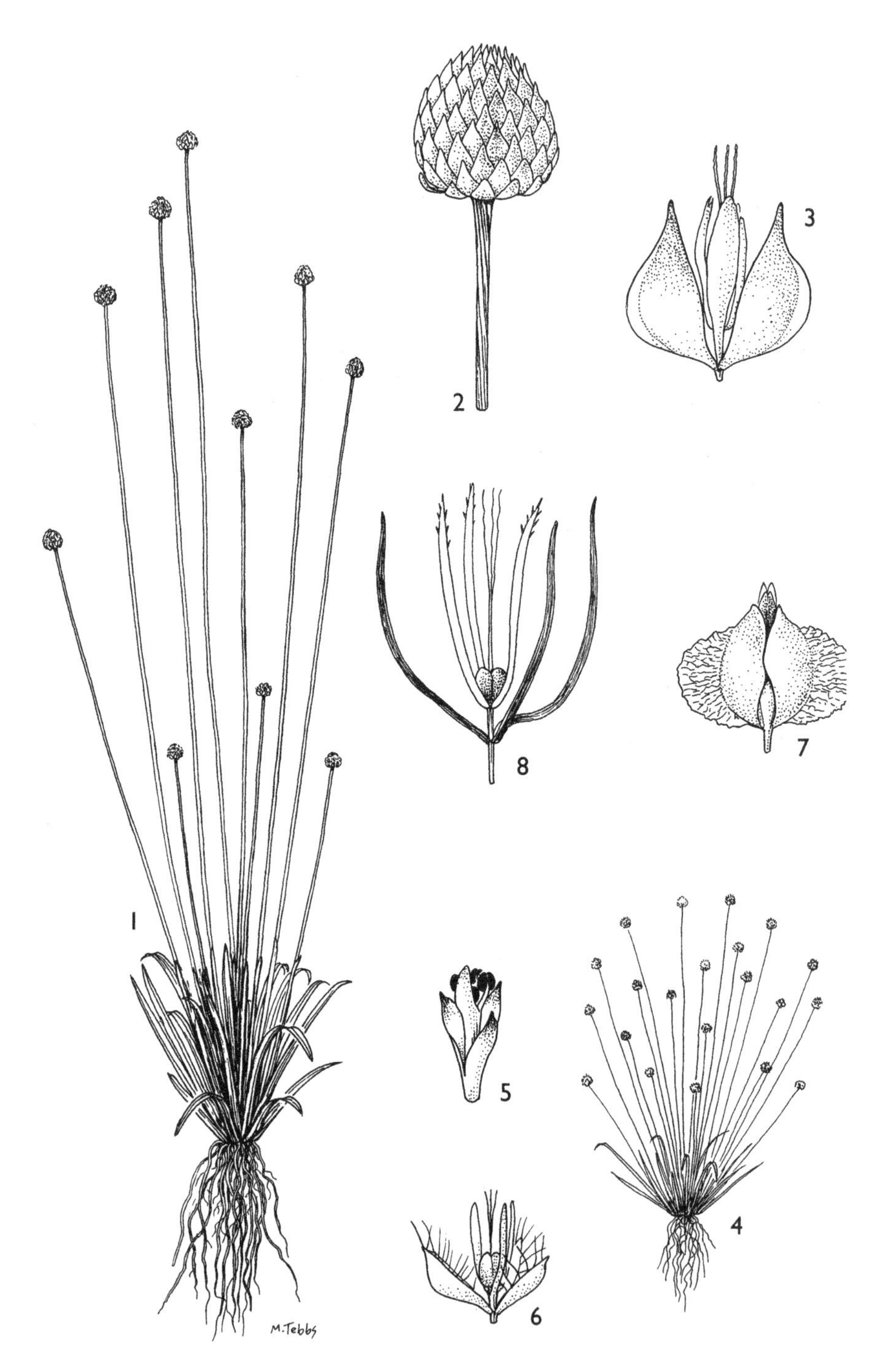

FIG. 3. *ERIOCAULON BUCHANANII* — **1**, habit, × $^2/_3$; **2**, capitulum, × 4; **3**, female flower, × 14. *ERIOCAULON ABYSSINICUM* — **4**, habit, × $^2/_3$; **5**, male flower, × 14; **6**, female flower, × 14. *ERIOCAULON MUTATUM* — **7**, female flower, × 14. *ERIOCAULON BONGENSE* — **8**, female flower, × 14. 1–3, from *Faden et al.* 96/97 & 96/482; 4–6, from *Faden et al.* 96/396; 7, from *Milne-Redhead & Taylor* 9890; 8, from *Faden et al.* 96/329. Drawn by Margaret Tebbs.

traditionally been separated from the other species listed in the synonymy by the possession of a glabrous receptacle and glabrous sepals. In fact, a few hairs are usually present in one or both locations and are often more numerous and obvious. Whilst plants from the southern part of the range tend to be hairier, glabrous plants occur as far south as Natal and hairy ones north to Uganda. A division into species on this character is untenable. *E. abyssinicum* intergrades with *E. welwitschii* (see note under *E. welwitschii*).

E. abyssinicum is frequently confused with *E. mutatum* and *E. cinereum*, all three being very similar small dark-headed annuals which sometimes grow together in open situations with other small pioneer species. *E. cinereum* can be distinguished by its white anthers and oblong obtuse floral bracts. *E. mutatum* has shorter, broader leaves, less obvious involucral bracts and broader, less scarious floral bracts. On dissection, it can be immediately distinguished from *E. abyssinicum* by its dimerous flowers.

18. **E. welwitschii** *Rendle*, Cat. Afr. Pl. Welw. 2: 97 (1899); N.E. Br. in F.T.A. 8: 249 (1901); Ruhland in E.P. 4, 30: 102 (1903); H. Hess in Ber. Schweiz. Bot. Ges. 65: 270 (1955). Type: Angola, Pungo Andongo, between Lombe and Candumba, *Welwitsch* 2441 (BM, holo.!)

Small rosulate annual. Leaves pale green, acicular, 1–2(–5) cm. long, 0.2–1.0 mm. wide, fenestrate, soon dying back. Scapes very slender, numerous, radiating to form a dome-shaped mound, 1–12 cm. high, 0.2–0.3 mm. in diameter, 3–4-ribbed; sheaths shorter than the leaves, scarious, obliquely slit. Capitulum hemispherical to dome-shaped, echinate, 2.5–4 mm. diameter, scarious, whitish to grey, the upper bracts often darker grey; involucral bracts as wide as the capitulum and radiating across its base, lanceolate with sharply acuminate tips, pale; floral bracts narrowly lanceolate, acuminate-aristate, flushed pale grey; receptacle columnar, glabrous to villous; flowers trimerous, pallid or flushed pale grey, 0.8–1.0 mm. long. Male flowers: calyx spathate, the sepals free or ± connate, the tips acuminate; petals tiny, glandular; anthers black. Female flowers: sepals equal, linear-lanceolate, concave, acuminate, the margins glabrous or with a few short hairs; petals linear, tipped with a rudimentary gland. Seeds elliptic, 0.3 mm. long, brown, glossy, faintly reticulate.

TANZANIA. Dodoma District: 48 km. S. of Itigi station on Chunya road, 21 Apr. 1964, *Greenway & Polhill* 11683!; Mbeya District: Ruaha National Park, Magangwe, 19 May 1968, *Renvoize & Abdallah* 2293!; Songea District: ± 6.5 km. W. of Songea, 28 Apr. 1956, *Milne-Redhead & Taylor* 9847!

DISTR. **T** 4, 5, 7, 8; Angola, Zambia, Zimbabwe, Namibia and South Africa (Transvaal)

HAB. Wet sandy ground in open places in bushland and *Brachystegia* woodland; 900–1550 m.

SYN. *E. welwitschii* Rendle var. *pygmaeum* Rendle, Cat. Afr. Pl. Welw. 2: 98 (1899). Type: Angola, Huila, near Lopollo, *Welwitsch* 2444 (BM, holo.!, K, iso.!)

E. aristatum H. Hess in Ber. Schweiz. Bot. Ges. 65: 163, t. 9/5, fig. 11 on p. 160 (1955). Type: Angola, Humpata District, *Hess* 52/1755a (Z, holo.!)

NOTE. *E. welwitschii* and *E. abyssinicum* represent different facets of a single intergrading complex, and consequently intermediates occur between them. The receptacle of both taxa varies from glabrous to villous, and like *E. abyssinicum*, the female sepals of *E. welwitschii* may be either glabrous or ciliate. *E. welwitschii* is distinguished by the combination of generally paler and narrower bracts and perianth parts, the tips of which are very narrowly pointed, imparting an echinulate appearance to the capitulum. The seed also tends to be a little smaller and more narrowly ellipsoid than in *E. abyssinicum*.

19. **E. zambesiense** *Ruhland* in E.J. 27: 75 (1899). Types: Malawi, Shire Highlands, *Buchanan* & Mt. Zomba, *Whyte* (both B, syn., K, isosyn.!)

Rosulate perennial from a short rhizome. Leaves linear, 5–8 cm. long, 3–4 mm. wide, many-nerved, spongy, subacute. Scapes 5–15, flexuous, 10–35 cm. high, 5–6-ribbed; sheaths equalling the leaves, straight-sided, obliquely slit, the limb acute. Capitulum subglobose, slightly wider than long, 5.5–7 mm. diameter, black and white, the white-hairy petal tips intermingling with the darker bracts, sometimes viviparous; involucral bracts small, straw-coloured or flushed grey, cartilaginous, oblong, 1.7–2.0 mm. long, rounded, glabrous, reflexed at maturity; floral bracts narrowly cuneate or oblanceolate, grey, white-pilose towards the acute tip; receptacle villous; flowers trimerous, 2.1–2.5 mm. long. Male flowers: sepals blackish, oblong, keeled, unequally connate below into an infundibular spathe, the tips broadly rounded, white-pilose; petals shortly exserted from the calyx, slightly unequal, white-pilose at the tips, the longest 0.7–1.0 mm. long; anthers black. Female flowers wedge-shaped, strongly trigonous: sepals navicular, blackish, two 1.5–2.0 × 0.8–1.0 mm., the third narrower, all strongly keeled, deeply concave and gibbous, keel with a spongy wing, white-pilose around the tip, margins villous along the inside with tubercle-based hyaline hairs; petals and ovary sessile; petals narrowly oblanceolate, shortly exceeding the sepals, slightly unequal, white-pilose at the tips, hairy on the inner blade with hairs similar to those along the sepal-margins. Seeds ellipsoid, 0.7 mm. long, yellow or brown with close-set transverse lines of white papillae.

DISTR. Burundi and Malawi, expected in S. Tanzania at ± 2000 m.

NOTE. *E. zambesiense* is closely related to *E. inyangense* (see note under that species).

20. **E. inyangense** *Arw.* in Bot. Notis. 1934: 83, fig. 2, 3 (1934). Type: Zimbabwe, Inyanga, Niarawe stream, *Fries, Norlindh & Weimarck* 2478 (LD, holo., BM, iso.!)

Rosulate herb, sometimes developing a short branching rhizome. Leaves in a dense tuft, linear-subulate to lanceolate, 2.0–5.5 cm. long, 2–3(–5) mm. wide, outwardly curving, spongy, acute. Scapes up to 25, slender, 7–35 cm. high, 5–6(–7)-ribbed; sheaths equalling the leaves, loose, obliquely slit, the scarious limb splitting into 2–3 lobes. Capitulum dirty white, 4–5 mm. diameter, initially globose with a flat base, becoming slightly elongated and dome-shaped with intruded base at maturity; involucral bracts shorter than the capitulum width, 1.5–1.8 mm. long, obovate-oblong, rounded, light brown, coriaceous at base becoming scarious and greyish upwards; floral bracts broadly spathulate, cuspidate, incurving, scarious, pale grey-brown, thinly white-pilose on the widest part; receptacle columnar, villous; flowers 1.3–1.8 mm. long, trimerous, pedicellate. Male flowers: sepals partially connate into an infundibular spathe, the laterals concave, the median flat, grey, tips obtuse and white-pilose; petals subequal, glandular and white-pilose, the largest 0.5–0.7 mm. long, shortly exserted from the calyx; anthers brown. Female flowers: sepals subequal, grey, deeply concave, lightly keeled, narrowly winged or at least spongily thickened on the centre keel, white-pilose with short stout hairs on back and margins towards the tip, margins usually pilose inside with longer hyaline hairs, lateral sepals 1.0–1.3 mm. long, the median slightly narrower, tip subacute; petals subequal, linear-oblanceolate, glandular, white-pilose at the tips, pilose on the inner blade with longer hyaline hairs. Seeds elliptic, 0.5 mm. long, brown, transversely white-papillose.

KENYA. Trans-Nzoia District: Kitale, Saiwa Swamp National Park, 17 Mar. 1977, *Hooper & Townsend* 1435!; Elgeyo Escarpment, July–Sept. 1926, *Harger*!; Ravine District: Timboroa [Timberoa], Lake Narasha, 2 Sept. 1971, *Livingstone et al.* 71-2!

TANZANIA. Iringa District: Dagaba Highlands, Kilolo, 9 Feb. 1962, *Polhill & Paulo* 1410!; Rungwe District: NE. of Rungwe Mt., bank of Kiwera R., 25 Oct. 1947, *Brenan & Greenway*!; Songea District: 12 km. W. of Songea, by Kimarampaka stream, 7 Jan. 1956, *Milne-Redhead & Taylor* 8064!

DISTR. **K** 3; **T** 7, 8; Zambia and Zimbabwe

HAB. Marshy grassland and streamsides; 950–2700 m.

SYN. *E. katangaense* Kimpouni in Fragm. Fl. Geobot. 39: 337, fig. 11 (1994). Type: Zaire, Shaba, 6 km. E. of Kasumbalesa, Lubumbashi–Sakania road, *Schmitz* 6131 (BR, holo.!)
[*E. fulvum* sensu U.K.W.F., ed. 2: 303 (1994), *non* N.E. Br.]

NOTE. *E. inyangense* has a typical facies of small, greyish, slightly elongate capitula on slender scapes rising above a basal tuft of often outwardly curving leaves. It is closely related to *E. zambesiense* which has a similar perennial rhizomatous habit, but this can be distinguished by its slightly larger, more obviously white-hairy capitulum which is never longer than wide, and especially by the much broader spongy wing to the female sepals. *E. inyangense* has most frequently been confused with the tall slender form of *E. transvaalicum*, subsp. *hanningtonii*, but this is clearly annual with flatter, thinner leaves up to 5 mm. wide and also seldom occurs above 1500 m.

The type specimen has rather stouter scapes than usual with 7 ribs, but otherwise matches the rest of the material well. The number of ribs in the scape of *Eriocaulon* species often varies within narrow limits and is not in itself a character of specific significance.

E. katangaense is probably this species, but the type is very immature. It is described as annual in the protologue, but the type specimen has a perennial rootstock.

21. **E. transvaalicum** *N.E. Br.* in Fl. Cap. 7: 54 (1897); Ruhland in E.P. 4, 30: 81 (1903); H. Hess in Ber. Schweiz. Bot. Ges. 65: 155, t. 8/10, fig. on p. 148 (1955); Oberm. in F.S.A. 4(2): 17, fig. 3/3 (1985); U.K.W.F., ed. 2: 303 (1994), as '*transvaalianum*'; Lye in Fl. Ethiopia 6: 379 (1997). Type: South Africa, Transvaal, Bosveld, Boekenhoutskloof [Buchenhoutskloofsspruit], *Rehmann* 4787 (K, holo.!, BM!, Z, iso.)

Rosulate annual or short-lived perennial, sometimes forming colonies. Leaves broadly linear to ensiform, 2–10 cm. long, 2–6.5 mm. wide, flat, thin, fenestrate, subacute. Scapes up to 30, usually strict, 8–35 cm. high, 0.5–1.0 mm. in diameter, 5-ribbed; sheaths a little shorter than the leaves, obliquely slit, the limb obtuse or splitting into 3–4 dry brownish teeth. Capitulum globose, shiny black, greyish white or infrequently brownish grey, (3–)5–9 mm. wide; involucral bracts as wide as or shorter than the capitulum width, scarious, usually paler than the rest of the capitulum, sometimes yellowish, 1.6–2.2 × 1.0–1.6 mm., oblong to obovate, obtuse, glabrous, reflexed at maturity; floral bracts spathulate to angular-oblanceolate, 1.8–2.3 × 0.6–1.0 mm., dark grey, acute, acuminate or cuspidate, varying from ± glabrous to pilose with short spaced white hairs on the upper back; receptacle villous; flowers trimerous, 1.5–2.0 mm. long. Male flowers: sepals black, connate into an infundibular spathe deeply split on one side, lobes free at the tips, glabrous or thinly white-hairy above, the upper margin often denticulate; petals reduced, included within the calyx, glands small or absent; anthers black. Female flowers: sepals subequal, dark grey, variable, navicular, narrowly lanceolate-falcate to ovate, winged on the centre keel or at least slightly thickened there, narrowed to the base, acute to sharply acuminate, glabrous or thinly white-pilose upwards, two 1.2–1.6 × 0.2–0.8 mm., the third narrower and often not keeled, 1.0–1.5 × 0.2–0.6 mm.; petals unequal, linear with dark tips, glabrous or white-pilose at the tip, infrequently some longer hyaline hairs (up to 0.8 mm.) also on the upper blade, glands vestigial or absent, one petal 1.5–2.5 × 0.2–0.4 mm., the other two 1.3–1.8 × 0.2 mm. Seeds ellipsoid, 0.4–0.6 mm. long, yellowish brown or infrequently reddish brown with transverse lines of minute white papillae.

KEY TO THE SUBSPECIES OF *E. transvaalicum*

Capitulum black, ± glabrous . . . . a. subsp. *dembianense*
Capitulum pilose with stout thick white hairs:
  Scapes stout, stiff . . . . b. subsp. *tofieldifolium*
  Scapes slender, often flexuous . . . . c. subsp. *hanningtonii*

a. subsp. **dembianense** (*Chiov.*) *S.M. Phillips* in K.B. 52: 57, fig. 1C–D, map 1 (1997). Types: Ethiopia, Dembia, near Mt. Inceduba near Gondar, *Chiovenda* 1651 (FT, syn., K, isosyn.!) & Scinta valley near Asoso, *Chiovenda* 1912 & 2573 (both FT, syn.)

Leaves ensiform, 2–10 cm. long, 2–5 mm. wide; scapes stiff, stout, 8–15(–20) cm. high, 0.6–0.8 mm. diameter. Capitulum shiny black or infrequently brownish grey, glabrous; floral bracts acuminate or cuspidate, usually ± glabrous, occasionally a few short white hairs on the back. Sepals of female flowers narrowly falcate, the upper margins denticulate, narrowly winged, tip cuspidate. Seed-surface tuberculate.

UGANDA. Masaka District: Lake Nabugabo, Aug. 1935, *Chandler* 1333!; Mengo District: 5 km. on Entebbe road, Aug. 1931, *Lab. Staff* 2/84! & Kampala, Kabaka's [King's] Lake, 4 Sept. 1935, *Chandler & Hancock* 35!

KENYA. Uasin Gishu District: Kaptagat, near Min. of Agric. station, Brockley Primary School, 7 Oct. 1982, *M.G. Gilbert & Mesfin* 6447!; Trans-Nzoia District: Kitale, 18 Sept. 1956, *Bogdan* 4299!

TANZANIA. Singida District: Iramba Plateau, Koimboi, 30 Apr. 1962, *Polhill & Paulo* 2263!; Mbeya District: Ruaha National Park, Magangwe, 19 May 1968, *Renvoize & Abdhallah* 2294!; Songea District: Kimarampaka, 28 Apr. 1956, *Milne-Redhead & Taylor* 9945!

DISTR. **U** 1, 4; **K** 3; **T** 4, 5, 7, 8; Zaire, Ethiopia and Zambia

SYN. *E. dembianense* Chiov. in Ann. Bot. Roma 9: 148 (1911)
E. sp. A sensu U.K.W.F.: 669 (1974)
[*E. transvaalicum* N.E. Br. var. *transvaalicum* sensu Lye in Fl. Ethiopia 6: 382, fig. 209.1/7 (1997), *non* N.E. Br. sensu stricto]

NOTE. This is the commonest form in northern and eastern tropical Africa. Subsp. *transvaalicum* occurs only in South Africa. It has the same black shiny capitula on stout, stiff scapes, but differs in details of sepal shape being ovate, deeply concave, with smooth margins and an acute tip and microscopic seed-surface patterning being ruminate.

b. subsp. **tofieldifolium** (*Schinz*) *S.M. Phillips* in K.B. 52: 58, fig. 2A–B, map 1 (1997). Type: Namibia, Hereroland, Waterberg, *Dinter* 378 (Z, holo.)

Leaves ensiform, 4–7 cm. long, 2.5–5 mm. wide. Scapes robust, strict, 10–20 cm. high, ± 1 mm. wide. Capitulum 6–9 mm. diameter, greyish white; floral bracts white-pilose above middle. Sepals of female flowers variable, ovate to lanceolate, wing narrow to broad with a toothed margin.

TANZANIA. Iringa District: Mkwawa [Mkawawa], 6 Nov. 1970, *Greenway & Kanuri* 14661!; Mbeya District: 10 km. SW. of Mapogoro, 29 Sept. 1970, *Thulin & Mhoro* 1262!

DISTR. **T** 7; Namibia and South Africa (Transvaal)

SYN. *E. tofieldifolium* Schinz in Bull. Herb. Boiss., sér. 2, 1: 779 (1901); H. Hess in Ber. Schweiz. Bot. Ges. 65: 266, fig. 5 on p. 265 (1955); Friedr.-Holzh. & Roessler in Prodr. Fl. SW.-Afr. 159: 2 (1967)

NOTE. The Tanzanian population differs from the southern African only in the presence of long hyaline hairs on the petals in the female flowers. It has much stouter, stiffer scapes than subsp. *hanningtonii*, which has similar white-hairy capitula.

c. subsp. **hanningtonii** (*N.E. Br.*) *S.M. Phillips* in K.B. 52: 58, fig. 2C–D, map 1 (1997). Type: Tanzania, Morogoro District, Kwa Chiropa, June 1803, *Hannington* (K, holo.!)

Leaves broadly linear, up to 10 cm. long, (2–)4–6.5 cm. wide. Scapes 9–35 cm. high, 0.5 mm. diameter, often flexuous. Capitulum 4–6.5 mm. diameter, greyish white; floral bracts white-pilose on the upper back. Sepals of the female flowers usually narrowly oblong-falcate, very narrowly winged, glabrous or a few white hairs near the tips.

KENYA. Northern Frontier Province: Lolokwi [Ol Lolokwe], cliffs opposite Subata Repeater Station, 15 Apr. 1979, *M.G. Gilbert* 5362!

TANZANIA. Moshi District: Mpololo, Aug. 1928, *Haarer* 1539!; Morogoro District: Kwa Chiropa, June 1803, *Hannington*!; Tunduru District: near Litungura, 6 June 1956, *Milne-Redhead & Taylor* 10660!

DISTR. **K** 1; **T** 2, 6, 8; Ethiopia, Mozambique and Zimbabwe

SYN. *E. hanningtonii* N.E. Br. in Fl. Trop. Afr. 8: 253 (1901); Ruhland in E.P. 4, 30: 74 (1903)
E. transvaalicum N.E. Br. var. *hanningtonii* (N.E. Br.) Meikle in K.B. 22: 142 (1968) & in F.W.T.A., ed. 2, 3: 63, fig. 336/11 (1968); Lye in Fl. Ethiopia 6: 382 (1997)

NOTE. Subsp. *hanningtonii* is distinguished by its slender, flexuous, often taller scapes and white-hairy capitula. The female sepals are more oblong and less distinctly winged than in subsp. *tofieldifolium.* It occurs mainly at low altitudes in eastern Africa.

Two specimens collected together at Thika (**K** 4; *Faden & Kabuye* 71/553 & *Kabuye* 370; *E. sp. B* sensu U.K.W.F., ed. 2: 303 (1994)) of a small *Eriocaulon* with grey-white capitula may belong here. The scapes are 3–4-ribbed and only ± 4 cm. high, and the flowers are also small (1.3–1.5 mm.) but agree in structure and seed morphology with subsp. *hanningtonii.* More collections are needed to determine their status.

DISTR. (of species as a whole) **U** 1, 4; **K** 1, 3, ?4; **T** 2, 4–8; Ethiopia, south to Namibia and South Africa (Transvaal, Natal and Orange Free State)
HAB. (of species as a whole) In mud of shallow water of streams, ditches and wet flushes in grassland, the leaves often submerged; sea-level–2350 m.

NOTE. (of species as a whole) A widespread and variable annual of wet muddy places, recognised by its broad, thin leaves and many black heads on stiff, relatively stout scapes. The female sepals vary considerably in width and shape, depending on how broad the wing is at the centre of the keel (often merely an indistinct thickening), and the degree to which the tips are acuminately extended. The female petals are frequently glabrous and eglandular, or may bear just a few short white hairs near the tip. There are also microscopic differences to the seed-surface patterning. These differences are discussed by S.M. Phillips (K.B. 52: 51–59 (1997)), who subdivides the species on the basis of floral and seed characters into 4 subspecies.

22. **E. selousii** *S.M. Phillips* in K.B. 52: 60, fig. 2E–F (1997). Type: Tanzania, Ulanga District, Selous Game Reserve, ± 8 km. S. of Ruaha camp, *Vollesen* in *M.R.C.* 2896 (K, holo.!).

Slender rosulate annual. Leaves broadly linear, 1.5–3.5 cm. long, 1–3 mm. wide, fenestrate, subacute. Scapes up to ± 15, 6–17 cm. high, much taller than the small basal rosette of leaves, 4–5-ribbed; sheaths equalling the leaves. Capitulum depressed-globose, 4–5 mm. diameter, brownish grey to dark grey with a slight sheen, glabrous, the bracts loose and untidy, the flowers visible between; involucral bracts as wide as the capitulum, yellowish grey, scarious, 2.0 mm. long, narrowly obovate, obtuse and sometimes minutely denticulate, reflexing at maturity; floral bracts resembling the involucral but narrower, cuneate-oblanceolate, shallowly incurving, obtuse and minutely denticulate, the inner acute; receptacle villous; flowers trimerous, 1.5–1.7 mm. long, glabrous. Male flowers: sepals connate into a thinly scarious, grey spathe with free tips, the spathe wide open, only the margins inturned, tips truncate-denticulate; petals very small, included within the calyx, eglandular; anthers black. Female flowers: sepals oblong with truncate-denticulate tips, laterals lightly keeled, the median similar but slightly narrower; all petals longer than the sepals, narrowly oblanceolate-oblong, eglandular, the median emarginate, laterals bidentate; ovary sessile. Seeds ellipsoid, 0.4 mm. long, brown with white-papillose reticulations.

TANZANIA. Ulanga District: Selous Game Reserve, ± 8 km. S. of Ruaha camp, 20 Oct. 1975, *Vollesen* in *M.R.C.* 2896!
DISTR. **T** 6; Malawi
HAB. Seepage in *Brachystegia* woodland; ± 275 m.

NOTE. The type formed part of a mixed gathering with *E. bongense* Ruhland. *E. selousii* is a segregate from the variable *E. transvaalicum* complex, distinguished mainly by its reticulate seed.

23. **E. stenophyllum** *R.E. Fr.*, Wiss. Ergebn. Schwed. Rhod.-Kongo-Exped.: 218, t. 16/1–2 (1916). Lectotype, chosen by S.M. Phillips in K.B. 53: ined. (1998): Zambia, Kali, *R.E. Fries* 634a (K, lecto.!)

Slender rosulate annual. Leaves delicate, narrowly subulate to filiform, extended into a long setaceous tip, 1.5–4.5 cm. long, 0.3–0.6 mm. wide, yellowish green. Scapes 2–6, slightly flexuous, 7–13 cm. high, 4–5-ribbed; sheaths inflated, usually a little shorter than the leaves, the mouth deeply slit, with acute limb. Capitulum hemispherical, 3–5 mm. diameter, pale buff with the bracts sometimes edged in black or occasionally mostly blackish; involucral bracts crateriform in 2–3 series, obovate with broadly rounded tip, 1.6–2.3 mm. long, cartilaginous; floral bracts oblong-cuneate, scarious, glabrous or with a few inconspicuous papillae on the back, tip broadly rounded with a central apiculum, upper margin denticulate; receptacle pilose; flowers trimerous, 1.3–2.0 mm. long, subsessile, pale or the perianth parts with dark tips. Male flowers: sepals ligulate, unequally and variably connate, the lateral lightly keeled, ± glabrous (sometimes a few inconspicuous papillae on the keel), tips emarginate-denticulate; petals rudimentary at the summit of a stipe which equals the calyx, glabrous and eglandular; anthers black. Female flowers: sepals resembling the male, two lateral falcate, spongy with large cells on the upper keel or sometimes broadened into a slight wing, keel sometimes with a few inconspicuous papillae, third sepal slightly smaller, straight; petals unequal, spongy, eglandular, ± glabrous (a few minute papillae on the upper margins), two narrowly lanceolate-oblong, emarginate, the median larger, lanceolate, 1.3–2.2 mm. long, bidentate; ovary raised on a stipe. Seeds plumply ellipsoid, 0.4 mm. long, reddish brown, closely transversely white-papillose.

TANZANIA. Songea District: Kwamponjore valley, 19 June 1956, *Milne-Redhead & Taylor* 10838! & 20 June 1956, *Milne-Redhead & Taylor* 10855!
DISTR. **T** 8; Zaire (Shaba), Zambia, Malawi and Botswana
HAB. Shallow water in boggy grassland; 1000 m.

SYN. *E. alaeum* Kimpouni in Fragm. Fl. Geobot. 39: 326, fig. 2 (1994). Type: Zaire, Shaba, Marungu Plateau, near Luonde, *Lisowski, Malaisse & Symoens* 6534 (BR, holo.!, BRVU, iso.)

NOTE. *E. stenophyllum* can be recognised by its thin subfiliform leaves, glabrous capitula with pale crateriform, cartilaginous involucral bracts and by its narrow sepals. The amount of blackish coloration in the capitulum appears to be very variable and of no taxonomic significance. *E. bicolor* Kimpouni is a dimerous offshoot from *E. stenophyllum* known only from the type collection from Zaire (Shaba). It also has wider wings on the female sepals. Reduction of the median carpel is sometimes seen in *Milne-Redhead & Taylor* 10838, and fully dimerous plants may also be found in S. Tanzania.

24. **E. afzelianum** *Körn.* in Linnaea 27: 680 (1861); N.E. Br. in F.T.A. 8: 250 (1901); Ruhland in E.P. 4, 30: 83 (1903); Meikle in F.W.T.A., ed. 2, 3: 62, fig. 336/10 (1968). Type: Sierra Leone, *Afzelius* (B, holo.)

Rosulate herb, probably a short-lived perennial. Leaves linear to linear-lanceolate, up to 7 cm. long, 1.5–4 mm. wide, thin, fenestrate, acute. Scapes up to 25, very slender, flexuous, 7–35 cm. high, sharply 4–5-ribbed; sheaths ± equalling the leaves, deeply obliquely slit with long acute limb. Capitulum 4–6 mm. diameter, subglobose to slightly dome-shaped, greyish with mealy white pubescence or sometimes dark grey, the involucral bracts yellowish, floral bracts neatly imbricate, the flowers hidden except for the largest female petal at maturity; involucral bracts broadly oblong to lanceolate-oblong with rounded tips, firm, often thinly mealy-pubescent, slightly smaller than the capitulum width and adpressed to it in several series; floral bracts narrowly cuneate-spathulate, scarious, uniformly grey or with pallid tips, incurved, white-pilose on the widest part, sharply acute; receptacle villous; flowers trimerous, shortly pedicellate. Male flowers 1.6–2 mm. long: sepals partially connate into a thin scarious spathe open on one side or sometimes almost free, oblong, the laterals concave, all white-pilose towards the obliquely truncate tips; petals included within the calyx, eglandular, white-pilose, the median 0.5–0.75 mm. long; anthers blackish. Female flowers 2–2.5 mm. long: sepals subequal, grey, narrowly oblanceolate-oblong, obtuse, thinly white-pilose towards the tips, the lateral two falcate, very lightly keeled, the median almost flat; petals raised on a short stipe, unequal, spongy, eglandular, white-ciliate on the upper margins, white-hairy also on both sides of the blade especially on the larger median petal, this longer than the sepals, elliptic with narrow base, 2–2.2 mm. long, the laterals smaller, narrowly oblanceolate. Seeds subglobose, ± 5 mm. long, reticulate (reticulations finally white).

TANZANIA. Mpanda District: 19 km. on Mpanda–Uvinza road, 14 May 1997, *Bidgood et al.* 3918!; Ufipa District: Kisumba–Mkowe road, 14 km. from Mkowe, 23 June 1996, *Phillips & Muasya* in *Faden et al.* 96/392!; Mbeya District: 67 km. on Tunduma–Sumbawanga road, 21 Apr. 1997, *Bidgood et al.* 3357!

DISTR. **T** 4, 7; W. Africa from Senegal to Chad and Central African Republic, Zambia and Malawi

HAB. Swampy grassland; 1100–1600 m.

SYN. *E. kouroussense* Lecomte in Bull. Soc. Bot. Fr. 55: 644 (1909). Types: Guinea, marsh of Kouroussa, *Pobéguin* 615 (P, syn., K, isosyn.!) & 616 (P, syn.)

NOTE. *E. afzelianum* is a widespread annual of variable habit, but well-grown specimens are rather tall with flexuous scapes and thin cellular leaves often over 5 cm. long. The southern populations in Tanzania, Zambia and Malawi have darker grey capitula than West African specimens, but there are no other differences.

Smaller specimens are very similar in facies to *E. transvaalicum* N.E. Br. subsp. *hanningtonii* (N.E. Br.) S.M. Phillips. On dissection *E. afzelianum* can be immediately distinguished by the narrow, subequal sepals of the female flowers, and by the unequal spongy, eglandular petals hairy on both sides.

25. **E. crassiusculum** *Lye* in Nordic Journ. Bot. 16: 63, fig. 1–4 (1996) & in Fl. Ethiopia 6: 382, fig. 209.2 (1997). Type: Ethiopia, Kefa, Kocha, ± 5 km. E. of Jimma along road to Addis Ababa, *Friis, Getachew Aweke, Rasmussen & Vollesen* 2067 (K, holo.!, C, iso.)

Rosulate herb from an abbreviated rootstock, probably a short-lived perennial. Leaves narrowly lanceolate, 2–8 cm. long, 2–3 mm. wide, light green, fenestrate, tapering to a hardened obtuse tip, often with a pore on the upper surface. Scapes 1–12, slender, up to 35 cm. high and much exceeding the basal leaf rosette, 6–7-ribbed; sheaths equalling or a little longer than the leaves, loose, slightly inflated towards the oblique mouth, the limb with a scarious acute tip. Capitulum depressed-globose, 5–7 mm. diameter, creamy-white sometimes with a grey tinge, the white-hairy flowers visible among the pallid bracts; involucral bracts in several series subequalling the capitulum width, pale straw-coloured, thinly cartilaginous, 2.0–2.6 mm. long, obovate-oblong, rounded becoming lacerate, slightly downturned but not strongly reflexing at maturity; floral bracts resembling the involucral in colour and texture, angular-oblanceolate to spathulate, the outer acute, the inner cuspidate and thinly white-pilose on the back; receptacle pilose; flowers trimerous, 1.5–2.4 mm. long, pallid. Male flowers: sepals free or connate only at the base, the two lateral keeled, ovate-oblong, densely white-pilose towards the rounded tips, median sepal narrower and ± flat; petals small, white-pilose with small glands, longest petal ± 0.8 mm. long; anthers black. Female flowers: sepals very unequal, two oblong-navicular, the keel broadly winged in the lower half, the wing abruptly ending at the midpoint, sometimes in a tooth, white-pilose above the wing to the obtuse tip, median sepal linear, unwinged; petals linear-oblong, subequal, white-pilose on the inner face and some hairs also on the outer face, eglandular or one or more petals with a small gland. Seeds broadly elliptic, 0.5 mm. long, brown with white-fringed transverse lines.

UGANDA. W. Nile District: near Koboko [Kobboko], *Eggeling* 1852!; Masaka District: SW. of Lake Nabugabo, Bugabo, 1 Feb. 1969, *Lye* 1827! & NW. side of Lake Nabugabo, 9 Oct. 1953, *Drummond & Hemsley* 4693!

DISTR. **U** 1, 4; Ethiopia

HAB. Open areas in the wet mud of swamps or in shallow, slow-flowing water, with sedges and often *Miscanthus*; 1100–1250 m.

26. **E. angustibracteum** *Kimpouni* in Fragm. Fl. Geobot. 39: 329, fig. 4 (1994). Type: Zaire, Shaba, Kisenge, *Duvigneaud* 3259 (BRLU, holo.!)

Rosulate annual. Leaves narrowly linear to linear-subulate, up to 10 cm. long, 1–3 mm. wide, thin, fenestrate, tapering to an acuminate, sometimes setaceously extended tip. Scapes up to 20, 6–28 cm. high, 5–7-ribbed; sheaths shorter than the leaves, loose, obliquely slit with an obtuse, often splitting limb. Capitulum globose, 4–6 mm. diameter, when young the blackish floral disc is surrounded by spreading pale involucral bracts, maturing grey and buff, the white-pilose flowers visible among the bracts; involucral bracts obovate to rotund with rounded tip, 1.8–2.1 mm. long, pale straw or greyish tinged, firmly scarious, reflexing at maturity; floral bracts narrowly cuneate or oblanceolate, grey-buff or blackish, glabrous or the inner thinly white-pilose, acute to shortly acuminate; receptacle pilose; flowers trimerous, 1.5–2.0 mm. long. Male flowers: sepals oblong, the lateral keeled, almost free or connate below the middle into a spathe, tips obtuse, white-pilose or glabrous; petals small, subequal, eglandular, white-pilose; stamens usually 3, rarely up to 6 (see note), epipetalous, black. Female flowers: sepals usually very unequal, the two lateral navicular, narrowed to the base, a fleshy wing on the centre keel, grey-buff or sepal-body grey with paler wing, wing-width variable, when broad the margin often toothed, white-pilose on the back above the wing or sometimes almost glabrous, tip acute to shortly acuminate; median sepal shorter, linear-oblanceolate or rarely only slightly smaller than the laterals, the tip sparsely white-pilose; petals subsessile, narrowly linear, eglandular, the tips white-pilose, the median a little larger. Seeds ellipsoid, 0.4 mm. long, brown with rows of white papillae.

TANZANIA. Iringa District: Mbeya–Iringa road, 7 km. N. of Lugoda [Brooke Bond] turn off, 11 June 1996, *Faden et al.* 96/155!; Songea District: Kwamponjore valley, 20 June 1956, *Milne-Redhead & Taylor* 10843A! & Hanga Farm, 27 June 1956, *Milne-Redhead & Taylor* 10913!
DISTR. **T** 7, 8; S. Zaire and Zambia
HAB. Marshy grassland and rice fields, the leaf rosette often submerged; 1000–1700 m.

NOTE. *E. angustibracteum* is the only species of *Eriocaulon* in the Flora area to possess only 3 stamens. The missing whorl of stamens between the petals is represented by minute stumps, or rarely one or more may bear a small anther. *Milne-Redhead & Taylor* 10913A completely lacks any black pigmentation in the capitulum. *Faden et al.* 96/156 is similar, but has 6 stamens and broader acute leaves. This specimen is unfortunately very immature.

27. **E. buchananii** *Ruhland* in E.J. 27: 83 (1899); N.E. Br. in F.T.A. 8: 247 (1901); H. Hess in Ber. Schweiz. Bot. Ges. 65: 145, t. 8/7–9, fig. 5–6 on p. 138 (1955). Types: Malawi, without precise locality, *Buchanan* 1168 (B, syn., BM, K, isosyn.!) & Tanganyika Plateau, Chitipa [Fort Hill], July 1896, *Whyte* (B, syn., K, isosyn.!)

Slender tufted annual. Leaves in a small basal rosette, pale bronzy green, much shorter than the scapes, broadly linear, 1–4.5 cm. long, 1–3 mm. wide, spongy, subacute. Scapes up to 50 or more in an erect bunch, slender, flexuous, 6–20 cm. high, 4–5-ribbed; sheaths equalling the leaves, crowded together, the limb scarious, acute. Capitulum globose to slightly ovoid, greyish brown or blackish, 3.5–5.5 mm. diameter, hard, glossy, completely glabrous, floral bracts loosely imbricate, the flowers scarcely visible among them; involucral bracts small, shorter than the capitulum width, light brown, firm, obovate-oblong, the tips rounded and incurved, reflexed at maturity; floral bracts angular-spathulate, ± flat, grey, glabrous, acute; receptacle villous; flowers trimerous, 1.2–1.5 mm. long, pedicellate. Male flowers: calyx spathate, thin, brownish grey, the free lobes subtruncate; petals very small, unequal, eglandular, the longest exserted from the calyx, 0.5 mm. long with a few small white papillae at the tip; anthers black. Female flowers: sepals very unequal, the two lateral deeply concave, frequently obtriangular, the keel gibbous with a broad spongy wing, narrowed to the base, tip acute, a few hyaline hairs often present inside, otherwise glabrous, third sepal linear to narrowly oblong; petals slightly unequal,

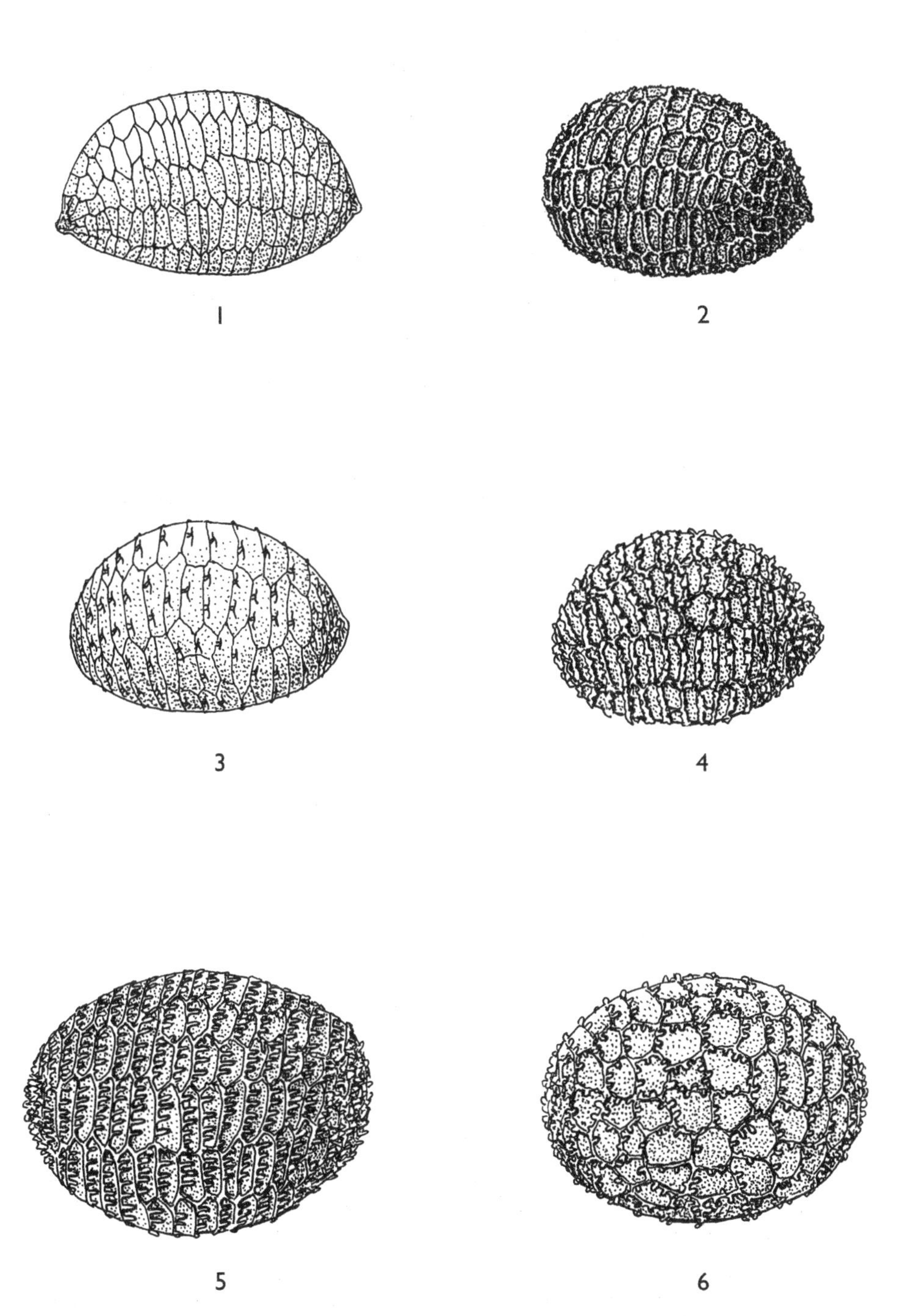

FIG. 4. Seeds of *Eriocaulon*, all × 100. **1**, *E. abyssinicum*; **2**, *E. nigrocapitatum*; **3**, *E. polhillii*; **4**, *E. elegantulum*; **5**, *E. buchananii*; **6**, *E. bongense*. 1, from *M.G. Gilbert* 4976; 2, from *Milne-Redhead & Taylor* 10841; 3, from *Greenway & Polhill* 11552; 4, from *Faulkner* 2174; 5, from *Buchanan* 1168; 6, from *E.A. Robinson* 5552. Drawn by Margaret Tebbs.

narrowly oblanceolate-oblong or the two smaller linear, eglandular or the two smaller with small glands or sometimes emarginate, greyish, the longest exserted from the calyx and often tipped with a few small white papillae. Seeds ellipsoid, 0.35 mm. long, light brown to reddish brown, with conspicuous transverse rows of white hair-like papillae. Fig. 3/1–3; fig. 4/5.

TANZANIA. Mbeya District: plains between Mbozi and Lake Rukwa, 10 May 1975, *Hepper & Field* 5484!; Songea District: ± 9.5 km. SW. of Songea, near Mtanda, 24 June 1956, *Milne-Redhead & Taylor* 10892! & N. of Songea, by Luhira R., 23 June 1956, *Milne-Redhead & Taylor* 10886!
DISTR. **T** 4, 7, 8; Burundi, Angola, Zambia, Malawi and Zimbabwe
HAB. Shallow water of pans, the wet sand or mud of recently flooded, drying hollows and in boggy patches dominated by *Loudetia* within *Hyparrhenia* grassland; 950–1900 m.

NOTE. The bunch of many, erect, slender scapes topped by hard, dark greyish globose capitula is characteristic of this species. The young heads are black (except for the paler involucral bracts), but frequently lighten to a greyish brown at maturity. The spongy gibbous "shoulders" of the female sepals are often visible among the loosely arranged floral bracts.

E. buchananii has very similar seeds with transverse rows of white papillae to *E. transvaalicum.* This is another annual with many scapes of globose blackish capitula, but is more robust with larger leaves and capitula, stiff, straight, diverging scapes, and it lacks the gibbous and spongily-winged female sepals so characteristic of *E. buchananii.*

E. fulvum is a closely related West African species, mainly distinguished by its pale capitula and obtuse floral bracts. *E. maculatum* and *E. strictum* also belong to this complex of erect annuals with very unequal female sepals and similar seeds, mostly with rows of white papillae. The boundaries between the individual species are not clear cut, and intermediate specimens may be encountered which are not easily assigned to species. *Milne-Redhead & Taylor* 10017 from Tanzania (Songea District) is very similar to *E. buchananii* but has rather small capitula (3.5 mm.) and floral bracts, perhaps due to introgression from *E. maculatum* Schinz.

28. **E. burttii** *S.M. Phillips* in K.B. 51: 636, fig. 6E–F (1996). Type: Tanzania, Dodoma District, Manyoni, Kazikazi, 17 May 1932, *B.D. Burtt* (K, holo.!)

Delicate tufted annual. Leaves few in a basal rosette, linear, 2–3.5 cm. long, 1–2.5 mm. wide, subacute. Scapes 1–5, 7–18 cm. high, erect in a narrow bunch, very slender and slightly flexuous; sheaths equalling the leaves. Capitulum globose, 3–5 mm. diameter, grey, hard, glabrous, slightly glossy; involucral bracts small, straw-coloured, obovate-oblong, rounded, 1.4–1.5 mm. long, tough, reflexing at maturity; floral bracts oblanceolate, flat, incurving, acute to cuspidate, numerous, the flowers visible between; receptacle villous; flowers 1.2–1.3 mm. long. Male flowers: sepals flushed dark grey, connate into an infundibular spathe with free, rounded, erose tips, glabrous, the laterals lightly keeled; petals small, included within the calyx, glabrous, the laterals with a rudimentary gland; anthers black. Female flowers: sepals dimorphic, grey, the laterals deeply concave, the keel gibbous with a broad spongy wing, narrowed to the base, tip subacute, glabrous or a few long hyaline hairs near the base inside, median sepal linear to oblanceolate, unwinged; petals slightly unequal, narrowly oblanceolate-oblong, laterals glandular, the median eglandular. Seeds ellipsoid, 0.3 mm. long, brown, glossy, faintly white-reticulate.

TANZANIA. Dodoma District: Manyoni, 44 km. S. of Itigi station on Chunya road, 20 Apr. 1964, *Greenway & Polhill* 11669! & Kazikazi, 17 May 1932, *B.D. Burtt*!; Mbeya District: Ruaha National Park, Magangwe Ranger Post, 9 May 1972, *A. Bjørnstad* 1652a!
DISTR. **T** 5, 7; not known elsewhere
HAB. Wet sandy places in open bushland; 1250–1400 m.

NOTE. A segregate from *E. buchananii* at the northern limit of the known range of that species, most easily distinguished by its more numerous narrower floral bracts and almost smooth seeds lacking white papillae.

29. **E. modicum** *S.M. Phillips* in K.B. 52: 68, fig. 6A–E, 7A–B (1997). Type: Tanzania, Songea District, without precise locality, *Milne-Redhead & Taylor* 10914A (K, holo.!)

Small annual. Leaves filiform, up to 1 cm. long, 0.1–0.3 mm. wide, spongy, finely acute. Scapes ± 10, 3.5–5 cm. high, 4-ribbed; sheaths almost equalling the leaves, loose, subinflated upwards. Capitulum depressed-globose, 2.5 mm. diameter, fuscous with slightly paler involucral bracts, the white-papillose petal-tips visible; involucral bracts as wide as the capitulum, crateriform, suborbicular, scarious, concave, rounded with crenulate upper margin; floral bracts obovate, obtuse to rounded with crenulate upper margin; receptacle shortly pilose; flowers trimerous, 1.0–1.2 mm. long, flushed grey upwards. Male flowers: sepals narrowly oblong, the two lateral falcate, keeled, median flat, unequally connate into a spathe or almost free, tips denticulate; petals small, glabrous, eglandular; anthers black. Female flowers: sepals very unequal, two laterals oblong-falcate, keeled, a spongy wing on the keel ± as wide as the sepal-body, tip acute, upper sepal-margins and upper wing-margin denticulate, a few white papillae below the tip; median sepal shorter, narrowly oblong, not winged, the tip denticulate; petals unequal, linear-oblong, eglandular, the tips entire, white-papillose. Seeds ellipsoid, 0.35 mm. long, brown, minutely punctate-papillose in transverse striae.

TANZANIA. Songea District: without precise locality, 27 June 1956, *Milne-Redhead & Taylor* 10914A!
DISTR. **T** 8; not known elsewhere
HAB. Muddy edge of drying pan; 1020 m.

NOTE. Known only from a single plant collected with *Milne-Redhead & Taylor* 10914 (*E. mutatum* var. *angustisepalum*).

30. **E. maculatum** *Schinz* in Bull. Herb. Boiss., sér. 2, 6: 709 (1906); Oberm. in F.S.A. 4(2): 14, fig. 2/7, t. 2/1 (1985). Type: South Africa, Transvaal, Blouberg, *Schlechter* 4651 (Z, holo, K!, PRE, SAM, iso.)

Small rosulate annual. Leaves linear, 0.5–1.5 cm. long, 1–2 mm. wide, minutely papillose, apiculate. Scapes up to ± 20, straight, 3–12 cm. high, 4-ribbed; sheaths equalling the leaves and similarly papillose, loose, the mouth 2–3-fid. Capitulum globose to slightly dome-shaped, 3–4 mm. diameter, brown-grey, floral bracts loose with the darker flowers visible between; involucral bracts as wide as the capitulum, pinkish brown with paler margins, scarious, obovate with rounded tips, reflexed at maturity; floral bracts resembling the involucral bracts in colour and texture, oblanceolate to spathulate, glabrous, subacute; receptacle cylindrical, villous; flowers trimerous, 1.0–1.5 mm. long, stipitate. Male flowers infundibular: calyx spathate with a truncate-erose margin, glabrous; petals included within the calyx, tiny, glabrous and eglandular; anthers black. Female flowers obtriangular: sepals very unequal, two deeply navicular, the keel gibbous with a pale, thick, spongy wing surrounding a thinner blackish central patch, margins flaring outwards with a few hyaline hairs inside, median sepal reduced, linear, sometimes as long as the other two but usually much shorter, sometimes vestigial; petals linear-oblanceolate, glabrous and eglandular, the median slightly larger. Seeds ellipsoid, 0.35 mm. long, reddish brown with a white reticulate patterning, glossy.

TANZANIA. Songea District: ± 6.5 km. W. of Songea, 28 Apr. 1956, *Milne-Redhead & Taylor* 9939!; Tunduru District: ± 1 km. E. of Songea District boundary, 6 June 1956, *Milne-Redhead & Taylor* 10655!
DISTR. **T** 8; southwards to South Africa (N. Transvaal)
HAB. Boggy grassland and the margins of drying pools on sandy soils; 900–1000 m.

NOTE. This small annual is very similar to *E. strictum*, but can be distinguished by its looser, toothed sheaths and broader leaves. The seed-coat patterning is also quite different. The papillae on the leaves and sheaths, although difficult to see except at high magnification, are a most unusual feature of this species.

31. **E. strictum** *Milne-Redh.* in Hook., Ic. Pl. 34, t. 3388 (1939). Type: Tanzania, Mafia I., Kilindoni, *Fitzgerald* 5213/3 (K, holo.!)

Slender tufted annual. Leaves narrowly linear, up to 3 cm. long and 1 mm. wide, 3-nerved, fenestrate, tapering to a filiform tip, soon dying back. Scapes 4–6, stiffly erect, 6–9 cm. high, 0.5 mm. diameter, 4-ribbed; sheaths loose, ± equalling the leaves, obliquely slit, the limb hyaline, entire, acute. Capitula subglobose, slightly longer than wide, 2.5–3 mm. diameter, grey with the darker flowers visible between the bracts; involucral bracts broadly obovate, subequalling the capitulum width, concave, the tip rounded, cuneate below, up to 1.3 mm. long and wide, straw-coloured; floral bracts obovate, lightly concave, tip rounded, cuneate below, ± 1.3 × 0.8 mm., glabrous, straw-coloured above; receptacle columnar, shortly pilose, the hairs 0.3 mm. long. flowers trimerous, ± 1 mm. long, pedicellate. Male flowers: sepals connate into a hyaline spathe; petals included, tiny; anthers black. Female flowers obtriangular: sepals two, deeply concave, the keel gibbous with a broad, pale wing surrounding a thinner blackish central patch, ± 1.1 × 0.8 mm., the third sepal vestigial, a tiny tooth; petals unequal, glabrous and eglandular, the tips blackish, lateral petals oblanceolate-oblong, ± 0.8 mm., the third oblanceolate, ± 1.2 mm. Seeds ellipsoid, 0.4 mm. long, with conspicuous transverse rows of white hair-like projections.

TANZANIA. Rufiji District: Mafia I., Kilindoni, 6 Aug. 1936, *Fitzgerald* 5213/3!
DISTR. **T** 6; not known elsewhere
HAB. Margins of drying pools; sea-level

NOTE. In its slender annual habit, and especially in the structure of the female flowers *E. strictum* is very close to *E. maculatum.* The two female sepals with a broad pale wing surrounding a central darker patch are almost identical, but the wing is less fleshy than in *E. maculatum* and the receptacle hairs are shorter. It is also readily distinguished by its narrower, almost filiform leaves and smooth sheaths with entire mouth. The floral bracts also lack the pinkish tinge of *E. maculatum.* The seed is like that of *E. buchananii,* but *E. strictum* is distinguished from that species by its stiff scapes, narrower leaves, smaller paler capitula, and by the absence of a third female sepal.

2. **MESANTHEMUM**

Körn. in Linnaea 27: 572 (1856); N.E. Br. in F.T.A. 8: 260 (1901); Ruhland in E.P. 4, 30: 117 (1903); Jacq.-Fél. in Bull. Soc. Bot. Fr. 94: 143–151 (1947); H. Hess in Ber. Schweiz. Bot. Ges. 65: 178 (1955); Meikle in F.W.T.A., ed. 2, 3: 64 (1968); Kimpouni in Fragm. Fl. Geobot. 39: 147–160 (1994)

Perennials from a stout tough rhizome or rarely slender annuals (annuals confined to West Africa); leaves and scapes frequently hairy. Leaves in a basal rosette, linear, spongy. Scapes few, unbranched, many-ribbed, arising from the leaf axils; sheaths obliquely slit, the limb acute. Capitulum globose or flattened; involucral bracts in several imbricate series, coriaceous with scarious margins, innermost sometimes radiating beyond the periphery of the white-hairy floral disc; floral bracts and flowers embedded in a woolly cushion of long receptacular hairs; floral bracts filiform with expanded hairy tips; flowers trimerous, pedicellate. Male flowers: sepals usually membranous, oblanceolate-oblong, concave, free or basally connate; petals connate into a fleshy infundibular tube with shallowly 3-lobed upper margin and basal stipe, petal-bases sometimes free, forming slits in petal-tube, 3 epipetalous glands towards the top within; stamens 6, arising from base of petal-tube; anthers yellowish, exserted above the corolla-rim at anthesis, later retracted into petal-tube; vestigial gynoecium present. Female flowers: sepals free, resembling the male sepals, caducous; petals free at the base around the ovary, connate above into a fleshy cylindrical tube or sometimes free to the middle, equalling or longer than the sepals, glabrous or villous both outside and within, 3 linear brownish

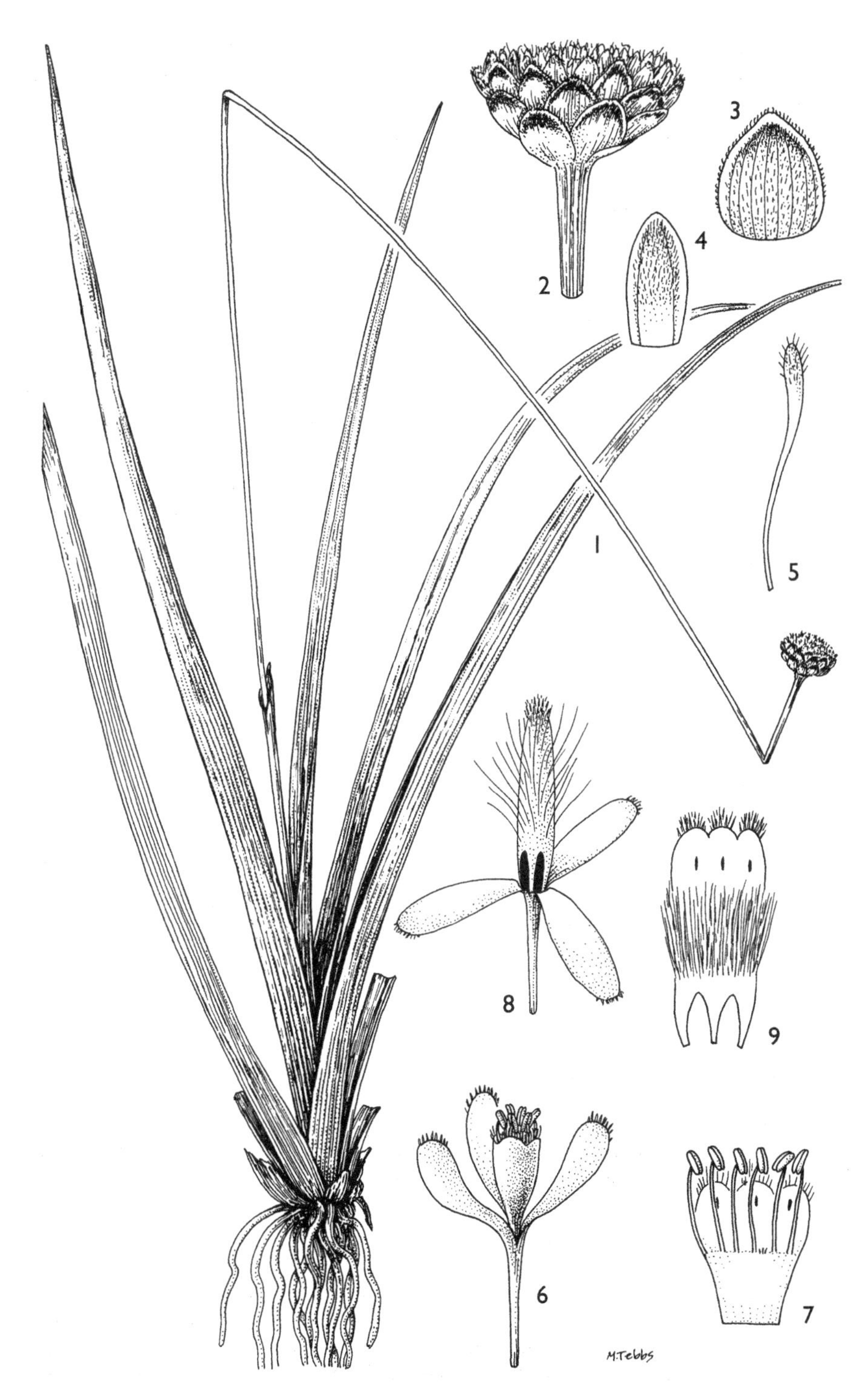

FIG. 5. *MESANTHEMUM RADICANS* — **1**, habit, × ⅔; **2**, capitulum, × 2; **3**, outer involucral bract, × 4; **4**, inner involucral bract, × 4; **5**, floral bract, × 14; **6**, male flower, × 14; **7**, inside view of male petal-tube with stamens, × 20; **8**, female flower, × 14; **9**, inside view of female petal-tube, × 20. All from *Norman* 54. Drawn by Margaret Tebbs..

glands inserted ± 2/3 up the inside, petal-tube thinner above level of gland insertion, densely white-pilose at tip. Seeds subglobose, brown, densely covered in white hair-like projections.

15 species in tropical Africa and 2 endemic in Madagascar.

A small genus with most species of restricted distribution, concentrated especially in West Africa and from S. Zaire to N. Zambia. Most are robust tussocky perennials, larger than all but the most vigorous perennial species of *Eriocaulon*. The single species known from the Flora area can be distinguished from *Eriocaulon* without dissection of the flowers by its cup-shaped involucre with the inner bracts extending beyond the floral disc.

M. radicans (*Benth.*) *Körn.* in Linnaea 27: 573 (1856); N.E. Br. in F.T.A. 8: 260 (1901); Ruhland in E.P. 4, 30: 119 (1903); Jacq.-Fél. in Bull. Soc. Bot. Fr. 94: 147 (1947); H. Hess in Ber. Schweiz. Bot. Ges. 65: 180 (1955); Meikle in F.W.T.A., ed. 2, 3: 64, fig. 339 (1968); F.P.U., ed. 2: 198 (1971); Kimpouni in Fragm. Fl. Geobot. 37: 134 (1992). Types: Sierra Leone, *Don* (K, syn., whereabouts uncertain) & Liberia, Grand Bassa, *Ansell* (K, syn.!) & Angola, *Curror* (K, syn.!)

Robust tussocky perennial from a short or elongating rhizome. Leaves clustered at the rhizome tip, 12–50 cm. long, (5–)10–15 mm. wide, scattered-pilose or glabrescent, rarely hirsute with patent hairs. Scapes tough, 30–60 cm. high; sheaths usually pilose, the limb tomentellous within. Capitulum creamy-white, flat-topped with a campanulate involucre, 10–15 mm. diameter; involucral bracts coriaceous with thinner margins, straw-coloured with greenish tips, patchily appressed-sericeous, outermost 3.0–3.6 mm. long, broadly ovate, rounded, innermost extending shortly beyond the floral disc, oblong, obtuse or subacute, white-hairy on the inner face; floral bracts capillary with a white-pilose subulate tip; receptacular hairs dark grey; flowers 2.2–2.8 mm. long. Male flowers: sepals free, pallid, broadly oblong, tips truncate and white-ciliate. Female flowers: sepals pallid, half as long to subequalling the petals, ovate-oblong, truncate-denticulate, glabrous or the tips white-ciliate in a central tuft; petals free near the base, central connate portion villous with long grey hairs outside and within, tips densely pilose with short white hairs. Fig. 5.

UGANDA. Masaka District: Sese Is., Bugala, Kalangala, 2 Mar. 1933, *A.S. Thomas* 934! & Lake Nabugabo, Aug. 1935, *Chandler* 1328! & S. of Luunga, Jubiya, 31 May 1971, *Katende* 945!
TANZANIA. Bukoba, June 1931, *Haarer* 2038! & Bukoba aerodrome, 21 June 1934, *Gillman* 74!; Rufiji District: Mafia I., near Kiombani [Kiombeni], 4 Oct. 1937, *Greenway* 5393!
DISTR. **U** 4; **T** 1, 6; Senegal to Cameroon, Congo, Zaire, Angola, Zambia and Mozambique
HAB. Swampy grassland, sphagnum bog, lake margins and wet coastal sand; sea-level–1200 m.

SYN. *Eriocaulon radicans* Benth. in Hook. & Benth., Fl. Nigrit.: 547 (1849)
E. guineense Steud., Syn. Pl. Glum. 2: 273 (1855). Type: Gabon, *Jardin* (P, holo.)
Mesanthemum erici-rosenii T.C.E. Fr. in R.E. Fr., Wiss. Ergebn. Schwed. Rhod.-Kongo-Exped.: 218, t. 16/4 (1916). Type: Zambia, Lake Bangweulu (Bangweolo), Mbawala I., *E. von Rosen* 806 (UPS, holo.!)

NOTE. The most widespread species of *Mesanthemum*, but with a distribution mainly in western Africa. It is known in East Africa only from the vicinity of Lake Victoria and on Mafia Is, but is also recorded from Lake Bangweulu in N. Zambia and from coastal Mozambique.

3. SYNGONANTHUS

Ruhland in Urban, Symb. Antill. 1: 487 (1900) & in E.P. 4, 30: 242–279 (1903); H. Hess in Ber. Schweiz. Bot. Ges. 65: 185 (1955); Meikle in F.W.T.A., ed. 2, 3: 67 (1968); Oberm. in F.S.A. 4(2): 19 (1985); Kimpouni, Lejoly & Lisowski in Seyani & Chikuni, Proc. XIII Plenary Meeting AETFAT, Malawi: 567–581 (1994); S.M. Phillips in K.B. 52: 73–89 (1997)

Annuals or perennials, usually hairy, hairs frequently glandular. Leaves in a basal rosette, linear to subulate or acicular, woolly in the axils. Scapes usually 3-ribbed; sheaths obliquely slit. Capitulum scarious, white or brown, pilose to almost glabrous; involucral bracts in several series, glabrous or softly ciliate on the margins, sometimes also on the back, the inner hyaline and subtending flowers; floral bracts absent; receptacle woolly with long hairs surrounding the flowers; flowers trimerous, pedicellate, the pedicels villous. Male flowers: sepals lanceolate, lightly keeled, connate below the middle into an infundibular tube with free lobes; petals completely connate into a membranous, subtruncate tube; stamens 3, filaments adnate to the inside of the corolla, extended above the corolla-rim and bearing 3 white or yellowish anthers; rudimentary gynoeceum at base of corolla-tube, 3-branched, the branches with swollen glandular tips; after anthesis the filaments collapsing inwards and the corolla-tube closing over the anthers. Female flowers: sepals free, otherwise resembling the male; petals delicate, usually pilose, connate near the tips, the clawed bases free, the small free apical lobes incurled; ovary 3-carpellate, the carpels protruding between the free petal-bases, style forming a hollow tube divided at the tip into 3 long stigmatic branches usually with 3 alternating swollen-tipped glandular appendages, these appendages entangling with the incurled petal-tips after anthesis to form a clavate structure. Seeds plumply ellipsoid to cylindrical, usually longitudinally white-striate.

About 200 species, mainly in South America; 14 species known from tropical and South Africa.
The above description applies only to the African species. In America the genus is much more diverse, sometimes with an elongate, branching stem, and with a greater range of inflorescence and floral morphology. In Africa *Syngonanthus* is a critical genus of closely related species, which are mostly difficult to distinguish from one another. In contrast to *Eriocaulon*, flower structure is very uniform, and specific delimitation rests mainly on small differences in vegetative and capitulum morphology. The scapes are frequently densely hairy below the capitulum, but appressed-hairy between the ribs or glabrescent lower down. Glandular-capitate hairs are always patent.
It is not usually necessary to dissect the flowers further than an inspection of the sepals of the female flowers. The tubercle-based hairs on the sepals often fall off during dissection, but their position can be seen by the remaining tubercles, especially along the sepal-margins. The presence or absence of swollen-tipped appendages alternating with the 3 stigmatic branches is important taxonomically, but is very difficult to see in the tiny flowers without good magnification. Only two African species lack these appendages, and will not often be found in the Flora area. It is only worth searching for this character if the specimen is a small annual with a capitulum not exceeding 4 mm. wide.

Perennials from a rhizome or rootstock; scapes 20–45 cm. high, or if less capitulum dark golden-brown:
- Capitulum golden-brown; scapes many, up to 25(–30) cm. high; leaves 1–3 cm. long, acicular, recurving, in dense woolly-centred rosettes; sheath-limb spathe-like ····· 1. *S. wahlbergii*
- Capitulum white with a yellow-brown involucre; scapes 1–10, 20–45 cm. high; leaves linear, up to 7 cm. long; sheath-limb straight ······························ 2. *S. angolensis*

Slender annuals; scapes up to 25 cm. high; capitulum white or white with a yellowish-brown involucre:
- Capitulum 5–6 mm. diameter; involucral bracts pallid; swollen-tipped appendages alternating with the 3 stigmas present ·························· 3. *S. longibracteatus*
- Capitulum 3.5–4.5 mm. diameter; involucral bracts pale yellow to golden-brown; swollen-tipped appendages alternating with the 3 stigmas absent ············ 4. *S. schlechteri*

1. **S. wahlbergii** (*Körn.*) *Ruhland* in E.P. 4, 30: 247 (1903); H. Hess in Ber. Schweiz. Bot. Ges. 65: 186, t. 9/11, fig. 2–4 on p. 198 (1955); Meikle in F.W.T.A., ed. 2, 3: 67 (1968); Oberm. in F.S.A. 4(2): 19, fig. 4 (1985); S.M. Phillips in K.B. 52: 75, fig. 1 (1997). Type: South Africa, Transvaal, Kaapse Hoop [Goda Hoppsudden], *Wahlberg* (S, holo.)

Small tussocky perennial, the leaf-rosettes conspicuously woolly in the centre, growing in clusters from a branching rhizome. Leaves numerous, acicular, 1–3 cm. long, 0.4–0.8 mm. wide, thinly appressed-pilose to subglabrous, the narrowly obtuse tips often curving inwards. Scapes several to many, flowering in succession, up to 25(–30) cm. high, 3-ribbed, pilose with patent glandular hairs especially immediately below the capitulum, some appressed pointed hairs also present; sheaths glandular-hairy, loose, the mouth inflated with open spathe-like limb. Capitulum golden or dark brown, 5–6 mm. diameter, the bracts and sepals concolorous with the white petal-tubes visible between; outer involucral bracts oblong, obtuse, 1.5–1.8 × 0.8–1.0 mm., the inner progressively longer and narrowly elliptic-oblong, glabrous, innermost 2.0–2.5 mm. long, subacute, ciliate on the margins; receptacle villous. Flowers 1.5 mm. long, brown; sepals of the female flowers lanceolate, margins pectinate-ciliate, usually glabrous on the back or a few hairs on the median sepal, tips subacute, often minutely denticulate; male sepals variably connate, sometimes almost free, subacute-denticulate, pilose at the base of the lobes. Seeds uniformly brown at first, becoming white-striate when wetted. Fig. 6/8–9.

UGANDA. Masaka District: Lake Nabugabo, 6 Oct. 1953, *Drummond & Hemsley* 4637! & June 1954, *Lind* 348!; Mengo District: Entebbe, Sept. 1905, *E. Brown* 334!

TANZANIA. Bukoba District: Bukoba, June 1931, *Haarer* 2046 pro parte (mixed with *S. schlechteri*)! & Uzongora, 1880, *C.T. Wilson* 140!

DISTR. **U** 4; **T** 1, 4; Nigeria, Central African Republic, Zaire, southwards to South Africa (Transvaal)

HAB. Boggy or muddy margins of lakes, rivers and streams, often with sphagnum, the leaf-rosettes sometimes submerged in shallow water; 1100–1200 m.

SYN. *Paepalanthus wahlbergii* Körn. in Mart., Fl. Bras. 3(1): 459 (1863); N.E. Br. in Fl. Cap. 7: 59 (1897) & in F.T.A. 8: 263 (1901)

Dupatya wahlbergii (Körn.) Kuntze., Rev. Gen. Pl. 2: 746 (1891)

Syngonanthus chevalieri Lecomte in Bull. Soc. Bot. Fr. 55: 597 (1909) Types: Central African Republic, sources of the Ndellé, *Chevalier* 6818 (P, syn., K, isosyn.!) & Ndoutta, Télé marsh, *Chevalier* 8348 (P, syn.)

Eriocaulon recurvifolium C.H. Wright in K.B. 1919: 264 (1919). Type: Zaire, Atené, *Vanderyst* 3133 pro parte (K, holo.!) (mixed with *S. schlechteri* Ruhland)

NOTE. A small perennial species forming clusters of woolly-centred rosettes in bogs and marshes. The golden-brown capitula on glandular scapes are distinctive. The leaves have markedly incurving tips in most specimens from Zambia southwards, but in more northern populations (*S. chevalieri*) the leaves simply curve outwards without upturning tips.

The specimen *Nutt* s.n., collected in 1896 between Lake Tanganyika and Lake Rukwa, and cited by N.E. Brown (1901) as being this species, is in fact a specimen of *S. angolensis*.

2. **S. angolensis** *H. Hess* in Ber. Schweiz. Bot. Ges. 65: 193, t. 9/10 & 13, fig. 7–8 on p. 198 (1955); S.M. Phillips in K.B. 52: 79, fig. 2E–F (1997). Type: Angola, Bié, Rio Luassinga, 60 km. E. of Menongue [Vila Serpa Pinto], *Hess* 52/2098 (Z, holo.)

Tufted perennial from a short rootstock. Leaves narrowly linear, loosely erect or curving outwards, 1.5–7 cm. long, 0.6–1.7 mm. wide, pilose on both surfaces especially when young, glabrescent, acute. Scapes up to ± 8, 20–45 cm. high, 3-ribbed, pilose with pointed hairs, densely so below the capitulum, appressed-pilose between the ribs lower down, glandular hairs absent or sparse; sheaths equalling or longer than the leaves, softly pilose to subglabrous, the limb scarcely inflated, rather stiffly erect, tapering to an acute tip. Capitulum 7–9.5 mm. diameter, ivory-white to cream-coloured with a brown base and brown inside towards the centre; involucral bracts mostly shorter than the capitulum width, the longest reaching the periphery, brown near the scape, pallid around the periphery, subacute to acute, the outermost narrowly ovate, 2.0–2.3 mm. long, glabrous, the inner narrowly elliptic-oblong, 3.0–3.7 mm. long, innermost shortly ciliate on the margins; receptacle villous. Flowers 2.0–2.5 mm. long; female sepals pallid, narrowly oblong, ciliate on the

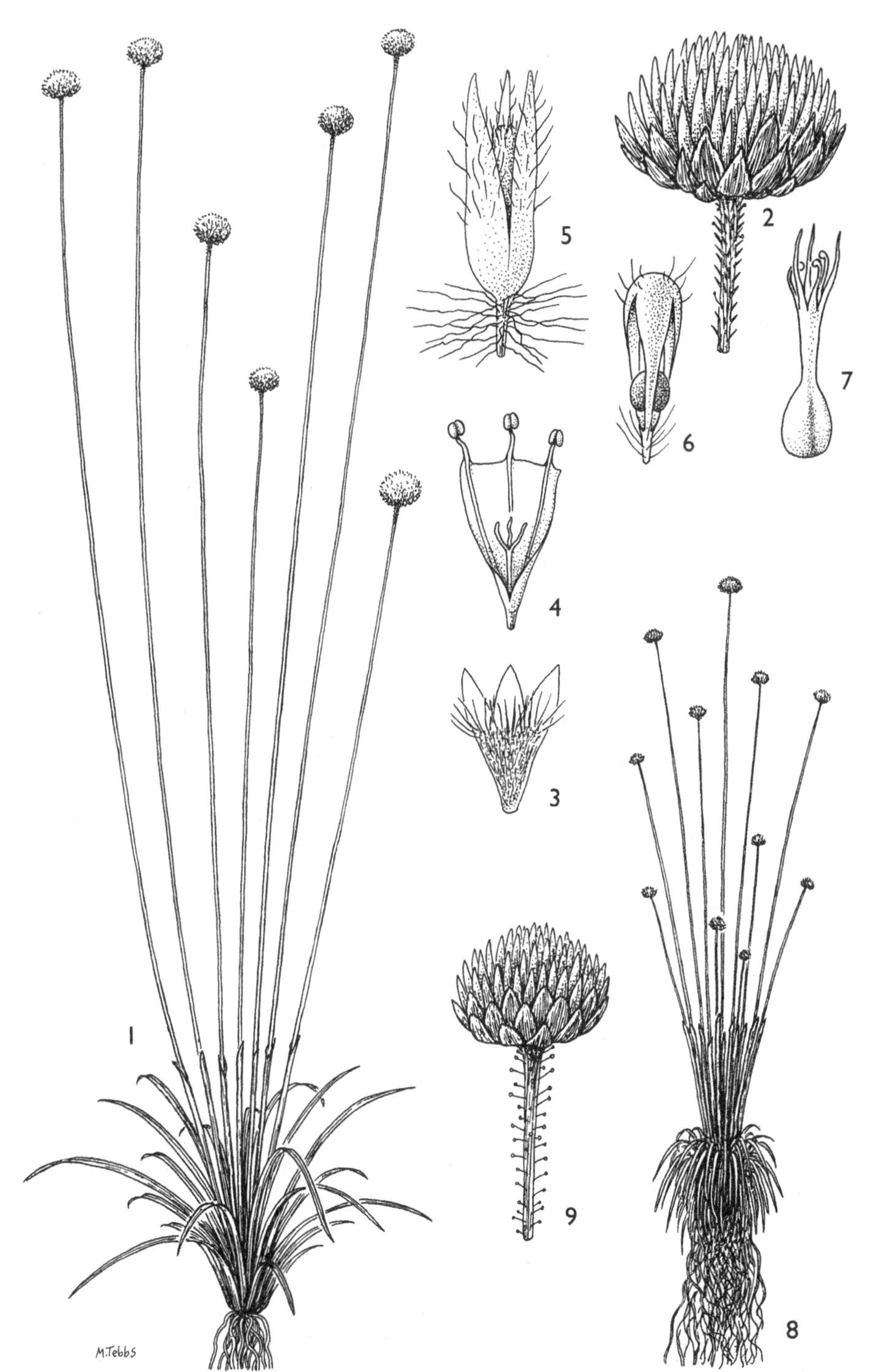

FIG. 6. *SYNGONANTHUS ANGOLENSIS* — **1**, habit, × $^2/_3$; **2**, capitulum, × 4; **3**, male flower, × 14; **4**, inside view of male petal-tube with stamens and vestigial ovary (not to scale); **5**, female flower, × 14; **6**, female petals, connate and shrunk inwards at top, and ovary, × 14; **7**, gynoecium, × 20. *SYNGONANTHUS WAHLBERGII* — **8**, habit, × $^2/_3$; **9**, capitulum, × 4. 1–5, from *Milne-Redhead & Taylor* 10847; 6, 7, from *Polhill & Paulo* 1524; 8, 9, from *Drummond & Hemsley* 4637. Drawn by Margaret Tebbs.

margins and back with fine caducous hairs, obtuse or acute; male sepals connate and dark brown below the middle, the free lobes pallid, obtuse-denticulate. Seeds ellipsoid, 0.5 mm. long, white-striate. Fig. 6/1–7.

TANZANIA. Iringa District: Dabaga Highlands, Kibengu, 17 Feb. 1962, *Polhill & Paulo* 1524!; Njombe District: Njombe–Kipengere road, ± 1.5 km. beyond Igosi, 26 Apr. 1970, *Wingfield* 595!; Songea District: 6.5 km. W. of Songea, 6 Feb. 1956, *Milne-Redhead & Taylor* 8704!
DISTR. **T** 4, 7, 8; Zaire, Zambia and Malawi
HAB. Marshy areas adjoining watercourses and boggy seepages on slopes in grassland; 950–2200 m.

NOTE. Most specimens of *S. angolensis* have pilose scapes with pointed hairs only, the hairs being particularly dense below the capitulum. However, short glandular hairs are occasionally also present below the capitulum intermixed with the longer pointed hairs, although they are never so obvious as in *S. longibracteatus.*

3. **S. longibracteatus** *Kimpouni* in B.J.B.B. 61: 339, fig. 2 (1991); S.M. Phillips in K.B. 52: 83, fig. 2D (1997). Type: Zaire, Shaba, Kundelungu Plateau, 8 km. NW. of Lualala, near source of Nungwe R., *Lisowski* 58176 (POZG, holo.!, BR, iso.)

Slender annual. Leaves many in a dense basal rosette, linear to subulate, 0.5–2.0 cm. long, 0.5–1.0 mm. wide, pilose with glandular and pointed hairs, subacute. Scapes up to ± 20 but often much fewer, 8–25 cm. high, 3-ribbed, pilose with patent glandular hairs intermixed with appressed pointed hairs; sheaths equalling or more often longer than the leaves, conspicuously glandular-hairy. Capitulum hemispherical to subglobose, 5–6 mm. diameter, shiny creamy-white or slightly yellowish; involucral bracts pale yellowish to light golden-brown, the outermost ovate-oblong and subacute, 1.4–1.7 × 0.6–0.8 mm., grading to oblong, the longest equalling the flowers, 2.3–2.8 × 0.7–0.9 mm., mostly glabrous, the innermost ciliate on the margins. Flowers 1.4–1.7 mm. long; female sepals lanceolate, acute, pectinate-ciliate along the margins, usually thinly hairy also on the centre of the back, occasionally the back glabrous; male sepals similar, acute, the basal connate portion light gold, the free lobes pallid. Seeds ellipsoid, 0.35–0.4 mm. long, white-striate.

TANZANIA. Dodoma District: Manyoni, Lambo ya Malengale, 7 July 1932, *B.D. Burtt* 3970!; Mbeya District: Madibira–Igawa track 14 km. SW. of Madibira, 12 June 1996, *Faden* et al. 96/187!; Songea District: ± 6.5 km. W. of Songea, 3 May 1956, *Milne-Redhead & Taylor* 9884!
DISTR. **T** 5, 7, 8; S. Zaire, Zambia and Zimbabwe
HAB. Waterlogged sandy soils and as a weed of rice, locally common; 950–1300 m.

NOTE. *S. longibracteatus* can be distinguished from similar annual species by its hairy scapes with a mixture of glandular and pointed hairs, and by the long, oblong-obtuse involucral bracts.
S. robinsonii Moldenke is another annual species to be expected in S. Tanzania near the Zambian border. It is known from N. Zambia and neighbouring parts of Zaire, where it frequently forms large colonies on damp sand among rocks or near waterfalls. It can be distinguished from *S. longibracteatus* by its scapes bearing only glandular hairs, and by its conspicuously hairy capitula. The involucral bracts are short, not exceeding half the capitulum width, so the capitulum has a brown base with a white marginal band. The seeds, which lack white striae, are very unusual in African *Syngonanthus*, and will confirm the identification.

4. **S. schlechteri** *Ruhland* in Schltr., Westafr. Kautsch.-Exped.: 272 (1901), *nom. nud.*, & in E.P. 4, 30: 247 (1903); H. Hess in Ber. Schweiz. Bot. Ges. 65: 190 (1955); Kimpouni in Fragm. Fl. Geobot. 37: 139, fig. 6 (1992); S.M. Phillips in K.B. 52: 86, fig. 2C (1997). Type: Zaire, Dolo, near Pool de Maleba [Stanley-Pool], *Schlechter* 12453 (B, holo., BM, BR, K!, P, Z, iso.)

Small tufted herb. leaves subfiliform, 1.5–2 cm. long, 0.3–0.6 mm. wide, ± glabrous. Scapes several to many, very slender, 5–15 cm. high, 3-ribbed, shortly pilose with patent glandular hairs and some pointed hairs; sheaths ± as long as the leaves, thinly pilose with glandular and pointed hairs, the limb straight, $^1/_4$–$^1/_3$ as long as the lower tubular portion. Capitulum white, appearing glabrous, 3.5–4.5 mm. diameter; involucral bracts slightly longer than the flowers, pallid, acute to narrowly obtuse, the outermost ovate, 1.4 mm. long, the inner elliptic, 2 mm. long, the innermost with a few short marginal cilia; receptacle villous. Flowers 1.0–1.4 mm. long; female sepals narrowly lanceolate-oblong, acute to subacute, glabrous on the back, pectinate-ciliate on the margins, the male sepals similar; style with 3 filiform stigmas but no alternating swollen-tipped appendages. Seeds ellipsoid, 0.3–0.4 mm. long, longitudinally white-striate.

TANZANIA. Bukoba, June 1931, *Haarer* 2046! pro parte (mixed with *S. wahlbergii*)
DISTR. **T** 1; Congo and Zaire
HAB. Marsh; ± 1150 m.

SYN. *Paepalanthus schlechteri* (Ruhland) Macbr. in Candollea 5: 348 (1934)

NOTE. This small species is the only species in East Africa lacking swollen-tipped appendages between the stigmas. The only other African species known to lack these appendages is *S. welwitschii* (Rendle) Ruhland, an even smaller species only 2–4 cm. high with few-flowered capitula up to 2.5 mm. wide. This occurs in Angola, Zambia and southern Zaire and is to be expected in southern Tanzania.

Subsp. *appendiculata* Kimpouni (in B.J.B.B. 61: 336, fig. 1 (1991)) has recently been described from southern Zaire and is also to be expected in upland parts of southern Tanzania. It differs mainly by possessing the swollen appendages between the 3 stigmas usually found in *Syngonanthus* but not in typical *S. schlechteri.*

S. schlechteri may hybridise with *S. wahlbergii* where both grow together. The Kew sheet of *Haarer* 2046 is a mixed collection of these two species with a specimen which appears to be intermediate. Hess (1955) mentions probable hybrids between *S. schlechteri* and *S. wahlbergii* from Angola.

4. PAEPALANTHUS

Kunth, Enum. Pl. 3: 498 (1841); N.E. Br. in F.T.A. 8: 262 (1901); Ruhland in E.P. 4, 30: 121–223 (1903); Rickett & Stafleu in Taxon 8: 232 (1959); Kimpouni in Fragm. Fl. Geobot., suppl. 2: 71–81 (1993), *nom. conserv.*

Variable herbs, annual or perennial, the stem short or elongate. Leaves thin to thick and coriaceous; capitula single or arranged in umbels. Capitulum villous; floral bracts present; flowers 2–3-merous; petals eglandular. Male flowers: sepals free except at the extreme base; petals connate into a glabrous infundibular tube bearing the stamens on its truncate upper margin; stamens as many as the petals, white; petal-tube finally involute and enclosing the stamens; a rudimentary pistil present. Female flowers: sepals free except at the extreme base and usually becoming rigid at maturity; petals free; style with glandular appendages alternating with the stigmas, these simple or bifid. Seeds variable, often with longitudinal ridges.

About 500 species, almost entirely confined to tropical America.

Paepalanthus is the largest genus in the family and is very heterogeneous, both vegetatively and in floral characters. Only two species occur in Africa, the widespread *P. lamarckii* and *P. pulvinatus* N.E. Br. in Sierra Leone.

P. lamarckii *Kunth,* Enum. Pl. 3: 506 (1841); Ruhland in E.P. 4, 30: 160 (1903); Lecomte in Bull. Bot. Soc. Fr. 55: 595 (1909); Milne-Redhead in K.B. 3: 472 (1948). Type: French Guiana, *Aublet* (P, holo.)

Small herb, with age developing an unbranched wiry stem clothed in old leaves and fibrous roots, the new leaves forming a loose rosette at the stem tip. Leaves linear to lanceolate, 2–3 cm. long, 1–3 mm. wide, spongy, scattered-pilose to glabrescent,

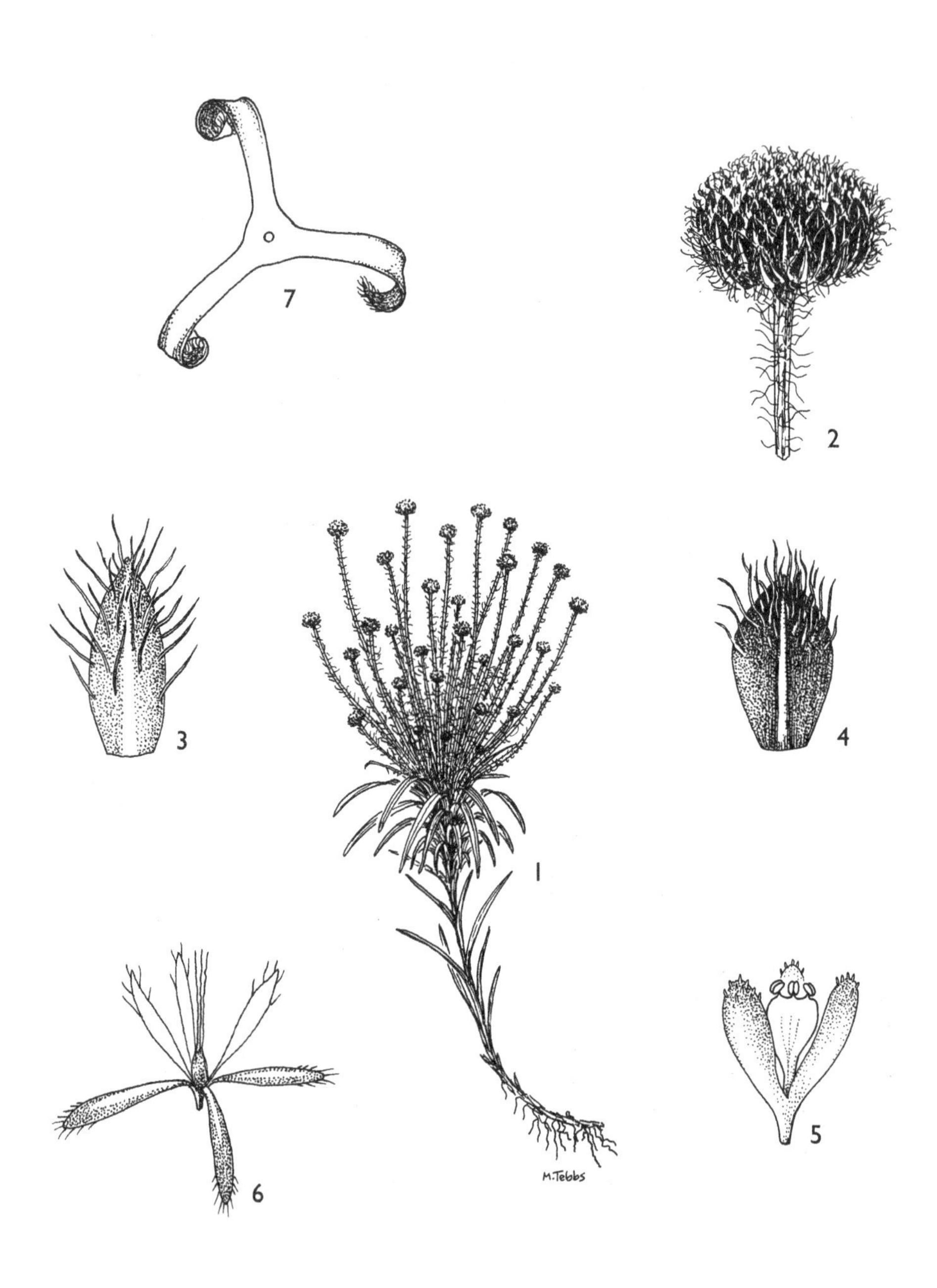

FIG. 7. *PAEPALANTHUS LAMARCKII* — **1**, habit, × $^2/_3$; **2**, capitulum, × 4; **3**, involucral bract, × 20; **4**, floral bract, × 20; **5**, male flower, × 20; **6**, female flower, × 20; **7**, recurved female calyx, after shedding petals and ovary, × 40. All from *Goyder et al.* 3081. Drawn by Margaret Tebbs.

the tip hardened, subacute. Scapes many in a terminal tuft, 2–7 cm. high, 3-ribbed; sheaths shorter than the leaves, loose with an acute limb. Capitulum 3–4 mm. diameter, subglobose with intruded base at maturity, grey, densely villous, the immature central flowers completely obscured by white hairs; involucral bracts lanceolate to lanceolate-oblong, 1.0–1.2 mm. long, firmly membranous, brownish grey with a paler central stripe, villous with coarse spreading hairs from the margins and back, acute; floral bracts angular-obovate, dark grey with paler central stripe, coarsely villous above the middle; flowers trimerous, 0.8–1.0 mm. long. Male flowers: sepals oblong-spathulate, concave, dark grey with paler central stripe, densely pilose at the subacute tips; petal-tube borne on a stipe; vestigial gynoecium represented by 3 elongate glands. Female flowers: sepals resembling the male at first; petals hyaline, equalling the sepals, narrowly oblong-spathulate, scattered-pilose; at maturity the sepals hardening, recurving and raising the petals and ovary to the capitulum surface. Seeds ellipsoid, 0.3 mm. long, light brown. Fig. 7.

TANZANIA. Mpanda District: 3 km. on Ugalla R. road from Mpanda–Uvinza road, 19 May 1997, *Bidgood et al.* 4049!; Rufiji District: Mafia I., Kilindoni, 6 Aug. 1936, *Fitzgerald* 5213/2!

DISTR. **T** 4, 6; scattered localities throughout West Africa, Zaire, N. Zambia and Madagascar, widespread in tropical America from British Honduras to Brazil and the West Indies

HAB. Around the drying margins of temporary pools and open places on damp sand and seepage areas; sea-level–1100 m.

SYN. *Eriocaulon fasciculatum* Lam., Encycl. Méth. Bot. 3: 276, t. 50/3 (1789), *non* Rottb. (1778). Type as for *P. lamarckii*

Eriocaulon lamarckii (Kunth) Steud., Syn. Pl. Glum. 2: 276 (1855)

Lasiolepis pilosa Boeck. in Flora 56: 91 (1873). Type: French Guiana, *Jelski* (B, holo.)

Dupatya lamarckii (Kunth) Kuntze, Rev. Gen. Pl. 2: 746 (1891)

NOTE. The method of seed dispersal is distinctive and unlike any method found in African *Eriocaulon*. The grey and white striped, hardened female sepals coil outwards, hence raising the petals and ovary to the capitulum surface where they are shed (Lecomte in Journ. de Bot., sér. 2, 1: 136, (1908)). The 3-armed, empty calyces with a central hole remain clinging to the capitulum, and are a good character for distinguishing this species from *Eriocaulon* without the need for dissection. The dry, brown, fibrous roots and villous scapes also serve to distinguish it from *Eriocaulon* without examining details of the tiny flowers.

INDEX TO ERIOCAULACEAE

GEOGRAPHICAL DIVISIONS OF THE FLORA

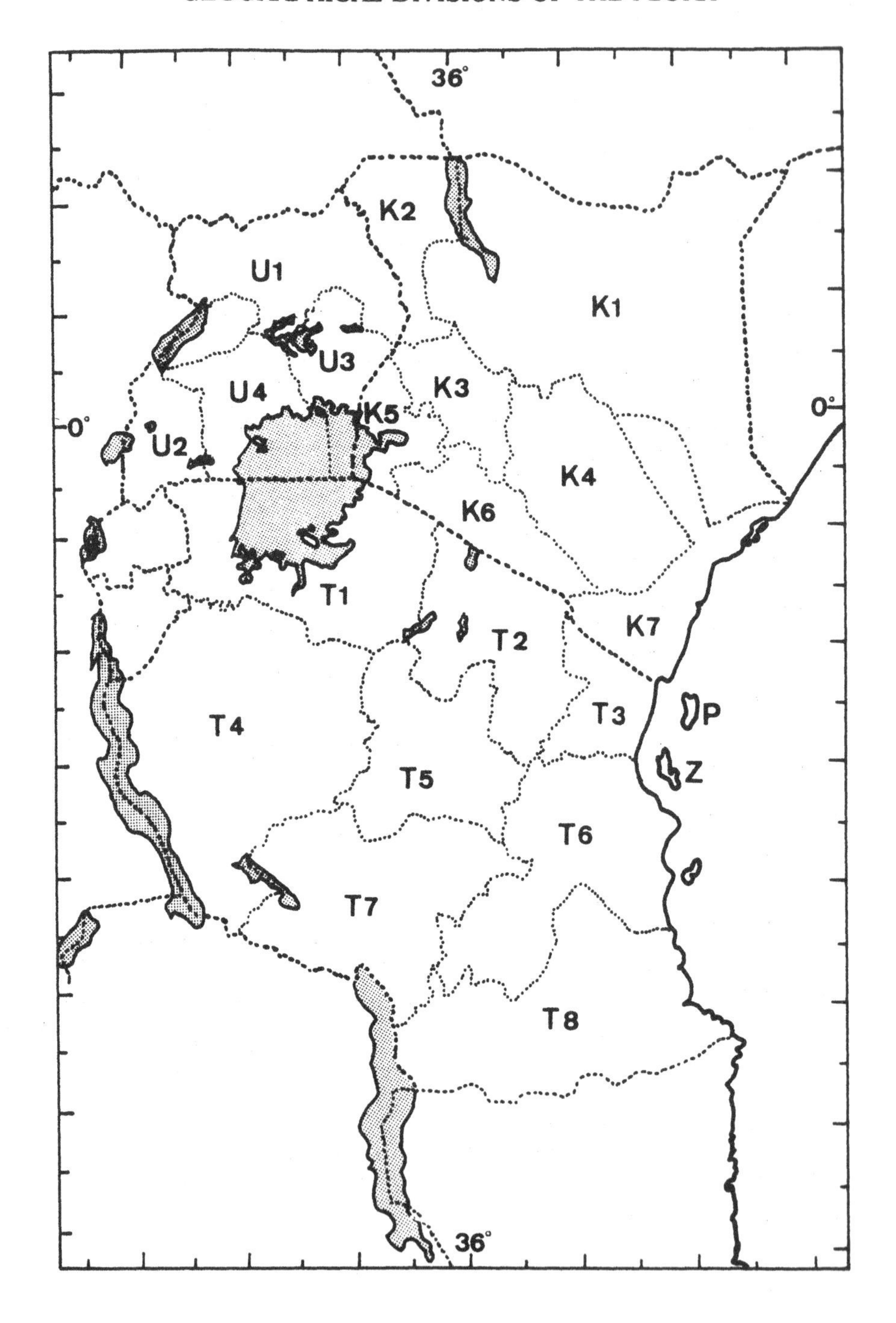

For Product Safety Concerns and Information please contact our EU representative GPSR@taylorandfrancis.com Taylor & Francis Verlag GmbH, Kaufingerstraße 24, 80331 München, Germany

T - #0010 - 090625 - C0 - 234/156/3 [5] - CB - 9789061913771 - Gloss Lamination